The Springer Series in Applied Machine Learning

Series Editor: Oge Marques

The Springer Series in Applied Machine Learning focuses on monographs, textbooks, edited volumes, and reference books that provide suitable content and educate the reader on how the theoretical approaches, algorithms, and techniques of machine learning can be applied to address real-world problems in a principled way.

The goal is a series of state-of-the-art books for the libraries of both machine learning scientists interested in applying the latest techniques to novel problems as well as practitioners, technologists, and researchers in various fields interested in leveraging the latest advances in machine learning for developing solutions that work for practical problems. The scope spans the breadth of machine learning and AI as it pertains to all application areas — both through books that address techniques specific to one application domain — and books that show the applicability of different types of machine learning methods to a wide array of domains.

Roshan Srivastav • Purna C. Nayak
Editors

Applications of Machine Learning in Hydroclimatology

Springer

Editors
Roshan Srivastav
Indian Institute of Technology Tirupati
Tirupati, Andhra Pradesh, India

Purna C. Nayak
National Institute of Hydrology
Roorkee, Uttarakhand, India

ISSN 2520-1298 ISSN 2520-1301 (electronic)
The Springer Series in Applied Machine Learning
ISBN 978-3-031-64405-4 ISBN 978-3-031-64403-0 (eBook)
https://doi.org/10.1007/978-3-031-64403-0

This Springer imprint is published by the registered company Springer Nature Switzerland AG
The registered company address is: Gewerbestrasse 11, 6330 Cham, Switzerland

Dedicated to the Almighty

Preface

The field of hydroclimatology is at a pivotal juncture, where traditional methodologies are increasingly being complemented and, in some cases, supplanted by advanced computational techniques. The advent of machine learning and artificial intelligence (AI) has opened new frontiers in our ability to model, predict, and manage water resources with unprecedented precision and efficiency. It is within this transformative context that *Application of Machine Learning in Hydroclimatology* has been conceived and developed.

This book is a comprehensive exploration of the immense potential that machine learning holds for addressing critical challenges in water resources management. It delves deeply into the application of AI techniques to unravel the complexities of hydrological systems, providing both researchers and practitioners with cutting-edge tools to enhance their work. The integration of machine learning into hydroclimatology represents a significant paradigm shift, enabling more accurate and reliable predictions of hydrometeorological extremes such as floods and droughts—phenomena that are becoming increasingly difficult to forecast due to the impacts of climate change.

In this volume, we examine a wide array of topics that are pivotal to modern hydroclimatology. We explore the modeling of hydrometeorological extremes, emphasizing the use of AI-driven methods to improve predictive accuracy. The book underscores the importance of analyzing non-stationarity and uncertainty in a rapidly evolving climate landscape, demonstrating how statistical and frequency analyses can enhance hydrological forecasts. Furthermore, it delves into the impacts of climate change on flood risks, drought occurrences, and reservoir operations, offering critical insights into how these phenomena affect water resource management.

To bridge the gap between theory and practice, the book includes numerous case studies showcasing effective mitigation measures for water-related challenges. These examples illustrate the practical application of machine learning techniques such as deep learning, reinforcement learning, and statistical downscaling in real-world scenarios. They highlight how AI can optimize decision-making and resource management, thereby improving our understanding of complex hydrological phenomena.

The diverse range of machine learning architectures presented in this book, including physics-guided models, data-driven techniques, and hybrid approaches, underscores the versatility and adaptability of AI in hydroclimatology. By leveraging these advanced techniques, researchers, practitioners, and policymakers are empowered to develop sustainable water management solutions that are both innovative and effective.

Application of Machine Learning in Hydroclimatology aims to bridge the gap between advanced AI technologies and hydrological science. It is our hope that this book will serve as a valuable resource, inspiring new approaches and solutions to some of the most pressing challenges in water resources management today. By bringing together the latest advancements in machine learning with the critical needs of hydroclimatology, we strive to contribute to a more sustainable and resilient future for water resources management.

Tirupati, Andhra Pradesh, India Roshan Srivastav

Acknowledgments

I am deeply grateful to all the contributing authors whose insightful chapters have enriched this edited volume. Your dedication, expertise, and willingness to share your knowledge have been invaluable in shaping this book. Each chapter reflects your commitment to advancing our understanding of *Applications of Machine Learning in Hydroclimatology*, and your contributions have significantly enhanced the quality and depth of this work. Thank you for your hard work, thorough research, and thoughtful writing.

I would like to extend my sincere appreciation to IIT Tirupati and NIH Roorkee for their unwavering support and encouragement throughout this project. Your resources, facilities, and academic environment have been crucial in bringing this book to fruition. The collaboration and support from both institutions have provided a solid foundation for this endeavor, enabling us to explore new perspectives and address complex challenges in our field.

A special thank you to my family and friends for their constant encouragement, patience, and understanding during the long hours spent on this book. Your love and support have been a source of strength and motivation for me. You have been my pillars of support, and I am deeply grateful for your presence in my life. Your belief in me and this project has been instrumental in its completion.

I also want to acknowledge the technical and administrative support from colleagues and staff at IIT Tirupati and NIH Roorkee. Your assistance in coordinating the contributions, managing deadlines, and ensuring the smooth progression of this project has been invaluable.

Finally, I would like to express my gratitude to all the reviewers and advisors who provided critical feedback and guidance. Your insights have helped refine this book and ensured its relevance and accuracy.

This book would not have been possible without the collaborative effort and support of everyone involved. Thank you all for making this journey a rewarding and fulfilling experience.

Contents

Applications of Physics-Guided Machine Learning Architectures in Hydrology

Prashant Istalkar, Akshay Kadu, and Basudev Biswal

Abstract Accurate prediction of different components of a hydrological cycle is of central interest to hydrologists for guiding water-resources-related policies. Over the years, various modelling frameworks have been developed and tested worldwide. These modelling frameworks, generally, are classified into three types: physics-based or process-based, black-box or data-driven and conceptual models. The physics-based models try to represent the catchment scale physical processes using the equations governed by mass, momentum and energy balance principles. The data-driven models, on the contrary, focus on mapping the input to the output by employing linear or non-linear transfer functions that seldom consider the underlying physical process. The conceptual modelling frameworks that lie in between the above two frameworks focus on representing the dominant hydrological processes in simplified mathematical forms. According to a few recent studies, deep-machine learning-based models that come under the category of data-driven models outperform the well-established conceptual hydrological models. These studies reported that the deep-learning models can better capture the information available in the model input over the traditional conceptual models. However, since the deep-learning-based models are physically inconsistent, their applicability under unforeseen scenarios may lead to significant ambiguity in the forecast. Taking into consideration the pros and cons of the data-driven models, a framework called physics-guided machine learning (PhyML) has been gaining popularity in recent years. It is built on the notion of incorporating the physical process understanding in the data-driven models. There are several ways of building the PhyML, such as providing an output of a physics-based model as input to an ML-based model, applying mass balance constrain in ML models, etc. The contrasting approach is also possible where a component of the conceptual hydrological model is simulated using ML. The current chapter discusses the recent applications of PhyML in the field of hydrology and compares the performance of a PhyML model with its simple ML variant for a case of dry-weather flow prediction.

P. Istalkar (✉) · A. Kadu · B. Biswal
Department of Civil Engineering, Indian Institute of Technology Bombay, Mumbai, India

R. Srivastav, P. C. Nayak (eds.), *Applications of Machine Learning in Hydroclimatology*, The Springer Series in Applied Machine Learning,
https://doi.org/10.1007/978-3-031-64403-0_1

1 Introduction

Hydrological science involves the study of the distribution and movement of water from the atmosphere to the ocean. The hydrological cycle consists of complex and interconnected hydrological processes that operate at different temporal and spatial scales (Blöschl & Sivapalan, 1995). It is essential to study the elements of the hydrological cycle for several water resources management applications such as flood forecasting, drought monitoring, climate change impact assessment, etc.(e.g., Kadu & Biswal, 2022; Nayak et al., 2022; Unnithan et al., 2024).

Since the early development of hydrological science, understanding how precipitation transforms into evapotranspiration and streamflow has been a central interest of hydrologists. The first kind of hydrological models were based on simple empirical relationship between precipitation, runoff and other physiological variables (Peel & McMahon, 2020; Todini, 2011). With the advancements in the understanding of catchment-scale hydrological processes and the computational power, several modelling theories were proposed and tested over the last two decades. These modelling frameworks are broadly classified into three groups: (1) physically-based or process-based, (2) conceptual and (3) black-box or data -driven.

In the year 1969, Freeze and Harlan proposed the physics-based distributed hydrological model framework to account for the spatial heterogeneity exhibited by landscape. This framework applies the laws of physics such as conservation of mass, energy and momentum at a grid scale to simulate the movement of water by solving the partial differential equations (Kampf & Burges, 2007). In the early 1980s, these models received lot of attention due to their ability to represent the spatial variability, more realistic simulation of runoff generation and climate variability. However, these models demand large data to obtain free parameters in calibration. In addition, they are computationally intensive, require significant amount of computing power and time to run. Due to these constraints, the use of physics-based hydrological modes was limited (Beven, 1989).

Conceptual hydrological models are simplified representation of real-world hydrological system. These models are based on a set of simplified equations that describe important hydrological processes. Conceptual hydrological models are widely used because of their structural simplicity and less data requirement compared to complex physics-based models (e.g., Bergström, 1992; Moore, 2007; Perrin et al., 2003). It is assumed that free parameters of conceptual hydrological model are abstract representation of catchment features. But in reality, unlike physics-based hydrological models, free parameters cannot be obtained using in-situ measurement and need to estimate using observed streamflow timeseries (Perrin et al., 2003). These models exhibit varying degrees of complexity, free parameters ranging from 2–3 to very complex model structure (Bergström, 1992; Biswal, 2016; Perrin et al., 2003). Despite their simplicity, conceptual hydrological models are very effective in predicting the behaviour of hydrological systems.

The black-box, also known as data-driven models, offered new approach in hydrological modelling. Unlike physics-based or conceptual models, these models map the relationship between input and output using empirical equations. These models use statistical or machine learning algorithms to learn from the data and create predictive models without the need for a prior knowledge of the physical process that governs the system (e.g., Guo et al., 2011; Hsu et al., 1995; Kratzert et al., 2018).

2 Machine Learning Algorithms Used in Hydrology

This section primarily focuses on history of most commonly used machine learning algorithms in hydrological applications. At present, several machine learning textbooks with detailed mathematical explanation are available. So, we limit our discussion of ML-based models in terms of hydrological interpretation.

2.1 Conventional Machine Learning Algorithms

The pioneering research in the field of ML was conducted by Turing (1950) and Rosenblatt (1958). However, ML algorithms received significant attention after the introduction of Artificial Neural Networks (ANNs). ANN mimics the biological neural networks of the human brain and is capable of learning complex relationship between input and output. The 1980s marked a significant breakthrough in the field of machine learning with the introduction of the effectiveness of backpropagation in training ANNs (Rumelhart et al., 1986). This discovery fuelled a surge of research centred around ANNs and their wide-ranging applications across numerous disciplines, including hydrology. For the first time, Halff et al. (1993) demonstrated the application of feed-forward ANN for streamflow prediction, utilizing rainfall inputs. In the same year, Buch et al. (1993) predicted the streamflow in the Himalayan glacier basin and reported its superiority over energy balance and multiple regression model. In successive years several studies applied ANN to model rainfall-runoff relationship (Hsu et al., 1995; Smith & Eli, 1995). One of the reasons ANN was successful in rainfall-runoff modelling is its ability to effectively capture the non-linearity in system. ANNs were able to produce promising results in streamflow prediction. This made them an attractive option for hydrologist looking for new approaches to model hydrological processes.

Similar to ANN, in the early 1990s, Support Vector Machines (SVMs) also became popular due to their ability to model non-linear relationship with high accuracy (Vapnik, 2000). SVMs work by mapping the input data into high-dimensional feature space and finding the hyperplane that best separates the data into different classes. This makes SVMs well-suited to modelling complex hydrological processes that involve multiple variable and non-linear relationship (Boser et al., 1992). So, SVMs were employed in several hydrological applications such as streamflow prediction (e.g., Ch et al., 2013; Guo et al., 2011), drought prediction

(Chiang & Tsai, 2011; Kolachian & Saghafian, 2021; Raghavendra & Deka, 2014) and groundwater modelling (e.g., Suryanarayana et al., 2014; Tang et al., 2019).

One of the limitations of ANN and SVM is interpretation of prediction. This limitation is overcome by decision trees (Jehanzaib et al., 2021). Decision trees use hierarchical structure to represent decisions and their possible consequences. Each internal node of the tree represents a decision based on the value of feature or attribute, and each leaf node represents a predicted outcome. Decision trees have been popular in hydrology due to ease of use and ability to variety of data types (Abraham et al., 2019, 2019; Friedl & Brodley, 1997; Pal & Mather, 2003; Ragettli et al., 2017). However, decision trees have a tendency to overfit the training data, especially when the dataset does not contain all the relevant features. This limitation is overcome by Random Forest (RF), by working with multiple decision trees at once and making it robust (Random forests, 2001). Random forest identifies patterns and relationship in hydrological data that are not easily identifiable using traditional statistical methods. RF has been used to identify the most important factors affecting water quality (Tyralis et al., 2019), groundwater recharge (Baudron et al., 2013) and most important factors affecting streamflow variability (Booker & Woods, 2014; Schoppa et al., 2020).

Along with these mentioned algorithms there are several other ML-based model used in the hydrology, such as K-Nearest Neighbours (Caraway et al., 2014; Prairie et al., 2006; Samaniego et al., 2008), Logistic regression (Lee & Kim, 2021; Li & Willems, 2019; Mair & El-Kadi, 2013; Ozdemir, 2011) and Gaussian Processes (Holman et al., 2014; Sun et al., 2014; Yang et al., 2018). All these algorithms are grouped in the category of supervised learning. Supervised machine learning algorithms are trained on labelled data, where the input data is associated with a known output. On the other hand, unsupervised machine learning algorithms are trained on unlabelled data, where goal is to find patterns in the data, e.g., K-means clustering. Unsupervised machine learning algorithms are powerful tool to find the pattern in complex data (Aytaç, 2020; Feng et al., 2019; Xu et al., 2018).

2.2 Deep Learning

It has been observed that for complex problems involving large datasets, conventional machine learning algorithms often do not perform well (Rahimzad et al., 2021). Moreover, hydrological data includes multiple variables making it difficult to analyse using traditional machine learning approaches. So, researchers used architectures with stacked multilayer neural networks termed as 'Deep Learning' (Dechter, 1986). These multiple layers process input features using non-linear units, and the output of the previous layer is fed to the next layer. The deep learning algorithm gets advantage over traditional machine learning algorithm in handling multiple features using stacked layers. A simple example of deep learning algorithm is recurrent ANNs, a special variant of ANNs (Schmidhuber, 2015).

Convolution Neural Networks (CNNs) is a type of ANN with special architecture design to handle grid-like data, such as images and videos (Lecun et al., 1998). They typically consist of several layers including convolutional layers, pooling layers and fully connected layers. This is the reason CNNs showed potential in classification and prediction problems (Ding et al., 2016; Kim & Song, 2020; Lin & Weng, 2018). However, the applicability of CNNs is limited compared to other deep learning algorithms due to the inability to handle sequential data. This limitation was overcome by Long Short-Term Memory (LSTM), a type of recurrent deep learning algorithm. In hydrological data, past information is important. For example, rainfall event that happened in the past 60 days may contribute to current-day streamflow. The LSTM is designed to effectively model the complex systems and long-term dependencies available in input features. For the first time, Kratzert et al. (2018) demonstrated the application of LSTM for rainfall-runoff modelling. In subsequent years, several studies tested the potential of LSTM in hydrological applications (Ayzel et al., 2021; Li et al., 2021; Song et al., 2022).

3 Process-Based vs Machine Learning Models

There has been a long-standing philosophical debate in hydrology over the use of black-box models versus process-based models. This debate centres around the question whether a model should be based on physical principles or should simply be a statistical representation of observed data. On one end, researchers argue that hydrological systems are complex and involve interaction between physical processes demanding understanding of underlying physics to model it accurately. Process-based models are based on physical laws and principles that govern hydrological processes, which can provide more mechanistic understanding of the system.

On the other hand, proponents of black-box model argue that machine learning-based models can often provide accurate prediction without requiring physics of underlying processes. These two extreme opinions are not something very new; they have existed since the early development of hydrological science. In the early 2000s, ANNs were a breakthrough in the field of machine learning and became popular in earth science applications. However, ANNs (or other existing models in the early 2000s) were physically inconsistent and that is one of the reasons for their limited use. Also, the traditional black-box models did not show any significant improvement over process-based model except the fact that it can model system without its prior knowledge, i.e., a single model can be used in several applications. Moreover, traditional machine learning models fail for unseen scenario, i.e., for event not observed during training. These limitations of traditional machine learning models are overcome by deep learning models such as LSTM. Recent studies showed that LSTM is superior to any existing traditional conceptual or process-based models in predicting daily streamflow (Kratzert et al., 2018, 2019b; Nearing et al., 2021). Based on these outcomes, authors claim that process-based models are unable to extract full information available in model input which is not the case for LSTM

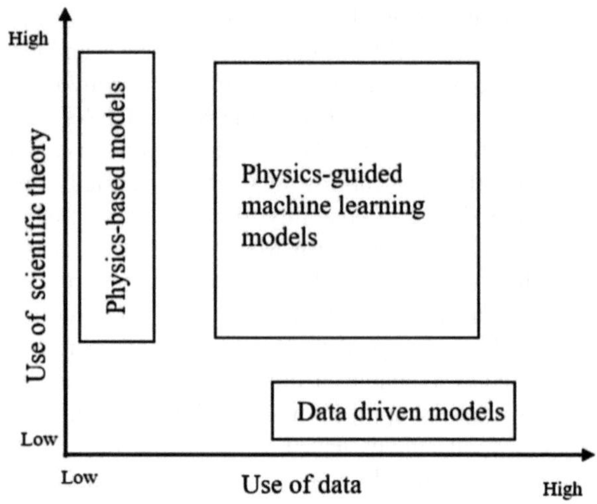

Fig. 1 Shows two extremes of hydrological modelling approaches, data-driven and physics-based models, with respect to data and scientific theories used. The physics-guided approach gets the benefits of both the approaches

(Nearing et al., 2021). Furthermore, studies reported the superiority of LSTM for ungagged basins showing its model structure robustness and ability of extrapolate for unseen event during training (Arsenault et al., 2023; Kratzert, Klotz, Herrnegger et al., 2019a; Zhang et al., 2022).

However, in any case, machine learning predictions are physically inconsistent. Moreover, "A better prediction without an understanding of the governing processes does not necessarily advance hydrological science." Also, the performance of machine learning models is primarily governed by the amount of data used in training (Ayzel & Heistermann, 2021). On the other hand, physics-based models are process-oriented and demand less data. The strength of both machine learning and process-based models is combined in physics-guided machine learning (e.g., Daw et al., 2017; Jia et al., 2021; Pawar et al., 2021; Xie et al., 2021). These models can overcome the limitation of two approaches by incorporating prior knowledge of physical principles in machine learning models (See Fig. 1).

4 Physics-Guided Machine Learning (PhyML) Models

In general, physics-guided machine learning (PhyML) architecture can be defined in two ways: incorporating physical information in ML models and incorporating ML in physics-based models.

4.1 Incorporating Physics in ML-Based Models

One of the ways to use physical information in machine learning is by adding regularization terms to the loss function of the machine learning algorithm that measure the deviation of model parameters from the physical constraints (Daw et al., 2017; Jia et al., 2021; Kaiser et al., 2018). These constraints are based on physical principles, such as conservation of mass, energy and momentum. The regularization terms penalise the deviation of the model parameters from physical constraints, leading to more accurate predictions (Karpatne et al., 2017). A study by Jia et al. (2021) predicted the temperature of lake with physics-informed recurrent neural network showing improvement in prediction by 20% with limited data during training. However, the regularization reduces the search space of machine learning model without modifying the core architecture keeping it still black-box. Recently, researchers attempted to modify the core architecture of ML algorithms to solve the specific problem. A study by Daw et al. (2020) modified the LSTM architecture by defining certain neurons with physical meanings, and found predictions are physically consistent. This helps in interpreting the architecture of ML with the predictions which is not possible in completely black-box models.

4.2 Incorporating ML in Physics-Based Models

ML models can be integrated with conceptual models in several ways. One of the approaches is to predict the important state variable of conceptual hydrological model using ML. For example, a study by Bhasme et al. (2022) used LSTM to predict the state variables of abcd model and ultimately monthly timeseries of evapotranspiration and streamflow. The study reported that the performance of PhyML model is better than physics-based model and ML model. Similarly, Khandelwal et al. (2020) integrated a SWAT model with LSTM to predict streamflow and found very similar outcome. A study by Xie et al. (2021) used synthetic rainfall samples for accurate prediction of flood peak, and proposed physics-informed LSTM showing its superiority over original LSTM. The another alternative way to improve the performance of conceptual model is to train the ML models for errors shown by conceptual models, and the same is demonstrated by Cho and Kim (2022).

5 Proposed Physics-Guided Machine Learning Models

It is true that the recent deep learning algorithms are smart enough to find the complex pattern between input and output. However, it is of prime importance to provide only significant information to ML models to train them effectively. In the current study, we discuss the physics-guided versions of ANN and LSTM proposed by Istalkar et al. (2023) for recession flow prediction, which utilises past streamflow information as model input. The study used the conceptual hydrological model, Power Law Regression (PLR), to decide the input to ANN and LSTM in their physics-guided versions. So, this study compared the performance of PLR, ANN, LSTM, physics-guided ANN (PhyANN), and physics-guided LSTM (PhyLSTM) in recession flow prediction.

5.1 *Power Law Regression (PLR)*

Recession flow is defined as continuously decreasing streamflow. Recession flow in a basin can be modelled using the relationship between storage and discharge, which can be expressed as a power-law equation (Eq. 1) with parameters k and α (Brutsaert & Nieber, 1977). The value of k depends on the wetness state of the basin (Biswal & Marani, 2014; Biswal & Nagesh Kumar, 2015), while α is related to the basin's properties (Patnaik et al., 2015). When the relationship between storage and discharge is not linear ($\alpha \neq 1$), the solution is given using Eq. 2 and can be used for prediction of recession curve.

$$-\frac{dQ}{dt} = kQ^{\alpha} \tag{1}$$

$$Q(t) = Q(0) \cdot [1 + (\alpha - 1)k \cdot Q(0)^{\alpha-1} \cdot t]^{\frac{1}{1-\alpha}} \tag{2}$$

$$k = k'_N \times Q_N^{-\lambda} \tag{3}$$

To estimate k, past discharge is used as a proxy for storage, and Eq. (3) is used to calculate k. In this equation, Q_N is the average past discharge from N to 1 day before the hydrograph peak. The value of N is selected using a method proposed by Reddyvaraprasad et al. (2020) that estimates the past average discharge representative of basin storage. The parameters of Eq. (3) are calculated using the least square linear regression, and the parameter k is estimated using Eq. (3). The recession flow is then predicted at a given time using Eq. (2). So, this employed this PLR model to predict 10 days of recession flow using past 120 days of streamflow information (Istalkar et al., 2023).

5.2 *ANN and LSTM*

An artificial neural network (ANN) is a type of machine learning algorithm that is loosely inspired by the structure and function of the human brain. It consists of multiple interconnected nodes, or "neurons," that work together to perform a specific task, such as image or speech recognition. The structure of an ANN typically consists of three types of layers: input, hidden and output. The input layer receives data, which is then processed by one or more hidden layers before being passed to the output layer, which produces a final result. Each neuron in an ANN receives input from one or more other neurons and produces an output based on a set of weighted connections between them. These weights are adjusted during the training phase of the ANN, where the network is exposed to a large set of labelled data to learn patterns and relationships within the data. In this study, we used a simple feedforward ANN network to predict 10 days of recession flow using the past 120 days of streamflow timeseries. Here, we employed 120 neurons in the input layer and 10 neurons in the output layer of the ANN. The number of neurons in the hidden layer was set to two-thirds of the sum of the input and output variables, as recommended by Han et al. (2007).

Long Short-Term Memory (LSTM) is a type of recurrent neural network (RNN) that is designed to address the vanishing gradient problem commonly associated with traditional RNNs (Hochreiter & Schmidhuber, 1997). The vanishing gradient problem occurs when the gradients in the backpropagation algorithm used to train the network become very small, making it difficult for the network to learn long-term dependencies. LSTM networks were first introduced in a 1997 paper by Hochreiter and Schmidhuber, and have since become a popular choice for a variety of sequence modelling tasks, including natural language processing, speech recognition and time series prediction. At the heart of an LSTM network are cells, which are responsible for maintaining a memory of the past inputs and outputs. The cells are designed to selectively forget or remember information over time, based on the input and the current state of the cell. Each cell in an LSTM network consists of three gates: the input gate, the forget gate and the output gate. The input gate determines which parts of the input to keep, while the forget gate determines which parts of the cell state to discard. The output gate determines which parts of the cell state should be output to the next layer. In addition to the three gates, LSTM cells also include a "memory cell" that stores the current state of the cell. This memory cell is modified by the input and forget gates, as well as by the cell's activation function.

One of the key advantages of LSTM networks is their ability to handle long-term dependencies. By selectively forgetting and remembering information over time, LSTM cells are able to maintain a memory of past inputs and outputs that can be used to inform future predictions. In this study, we used the past 120 days of streamflow information to predict 10 days of recession flow using LSTM. We employed encoder-decoder LSTM architecture in this study, as it was found to be better than the standard sequence-to-one LSTM architecture (Xiang et al., 2020).

5.3 Physics-Guided ANN (PhyANN), Physics-Guided LSTM (PhyLSTM) and Their Evaluation

As discussed previously, physics-guided ML is an approach that aims to incorporate the principles of physics into machine learning models to improve their performance and interpretability. One approach to PhyML is to build hybrid models that combine physics-based models with machine learning models. In these models, the final or intermediate output of a physics-based model can serve as input to machine learning models. In the proposed PhyANN and PhyLSTM models, the intermediate output of a conceptual model, PLR, is used as input along with peak flow to ANN and LSTM models. So, instead of providing the past 120 days of streamflow information to ANN and LSTM, we provided only streamflow information determined by Reddyvaraprasad et al. (2020) method. The idea here is to provide only sub-surface storage information as input to ANN and LSTM, which is responsible for recession flow generation.

Figure 2 illustrates prediction performance of PLR, ANN, LSTM, PhyANN and PhyLSTM models during the recession period across 252 study basins situated in the United States, in terms of NSE and logarithmic NSE (NSE_{log}). The median NSE values for PLR, ANN and LSTM stand at 0.71, 0.51 and 0.80, respectively. Here, the

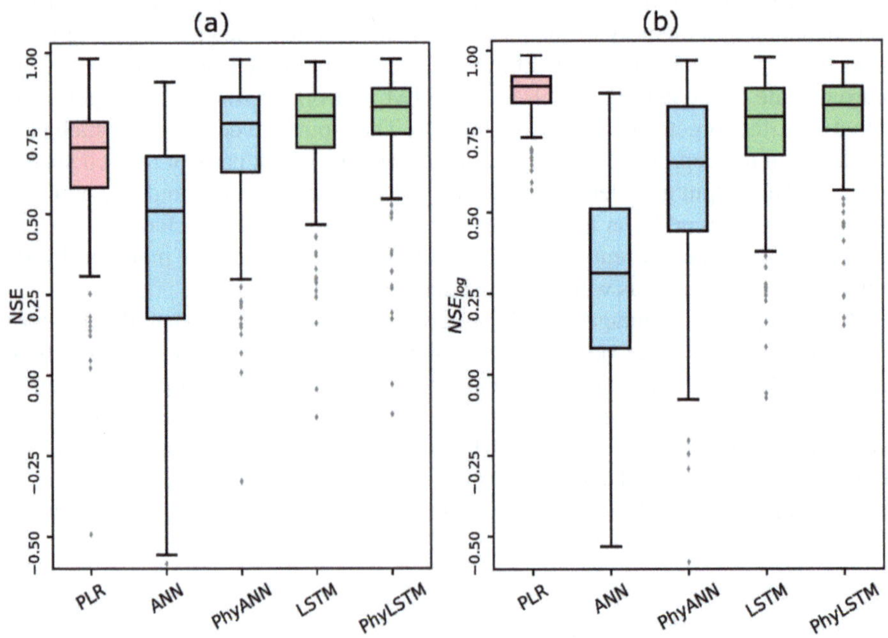

Fig. 2 Comparison of prediction performance of PLR, ANN, PhyANN and PhyLSTM models for 10-day recession flow in 252 basins in the United States, evaluated in terms of NSE and logarithmic NSE (NSE_{log})

best-performing model is PhyLSTM and least least-performing model is PLR. As studies reported, LSTM is able to extract the useful information efficiently compared to ANN and PLR. However, on providing the physics-informed to LSTM (PhyLSTM) it shows improvement. This improvement is quite high in case of ANN. This suggests that a simple machine learning model can get benefited more than advanced machine learning model on providing physics-informed input. Also, important point to note here is that the PhyANN performance is comparable to LSTM. This implies a simple machine learning model, like ANN, can perform as good as advanced machine learning model, like LSTM, on providing physics-informed inputs.

The median NSE_{log} values for PLR, ANN, PhyANN, LSTM and PhyLSTM are 0.89, 0.33, 0.66, 0.79 and 0.83, respectively. In case of NSE_{log}, a similar pattern can be seen where ANN and LSTM show the improvement in model performance on providing the physics-informed inputs. But the interesting point to note here is the conceptual PLR model is outperforming all models. This means the PLR model is able to capture the low flow dynamics better than ANN and LSTM. This shows that a conceptual model based on clear physical understanding can outperform the advanced machine learning models.

6 Summary

The aim of this study was to emphasize the importance of hydrological process understanding in the era of machine learning. In order to do so, we compared the predictive performance of a simple conceptual model (PLR) with two machine learning-based models (ANN and LSTM) and their corresponding physics-informed versions (PhyANN and PhyLSTM) for recession flow prediction, using only past streamflow data as input. The results indicated that the PLR model, which was built on a strong understanding of dominant processes, outperformed both the ANN and LSTM models. Additionally, the physics-informed machine learning models, PhyANN and PhyLSTM, showed better performance than their original versions. Although LSTM is an advanced machine learning algorithm, it sometimes struggles to extract relevant information from input data. However, its performance can be improved by incorporating physics-guided inputs. This suggests that physical process understanding can enhance the performance of recent advanced machine learning models. Furthermore, even a simple machine learning model like ANN can perform comparably to advanced models like LSTM when given physics-informed inputs. Overall, our study underscores the importance of process understanding in the use of machine learning-based models for hydrologic predictions. It also provides insights into the potential of physics-informed machine learning not only in hydrology but also in other fields.

References

Abraham, S., Huynh, C., & Vu, H. (2019). Classification of soils into hydrologic groups using machine learning. *Data, 5*, 2. https://doi.org/10.3390/data5010002

Arsenault, R., Martel, J.-L., Brunet, F., Brissette, F., & Mai, J. (2023). Continuous streamflow prediction in ungauged basins: Long short-term memory neural networks clearly outperform traditional hydrological models. *Hydrology and Earth System Sciences, 27*, 139–157. https://doi.org/10.5194/hess-27-139-2023

Aytaç, E. (2020). Unsupervised learning approach in defining the similarity of catchments: Hydrological response unit based k-means clustering, a demonstration on Western Black Sea Region of Turkey. *International Soil and Water Conservation Research, 8*, 321–331. https://doi.org/10.1016/j.iswcr.2020.05.002

Ayzel, G., & Heistermann, M. (2021). The effect of calibration data length on the performance of a conceptual hydrological model versus LSTM and GRU: A case study for six basins from the CAMELS dataset. *Computational Geosciences, 149*, 104708. https://doi.org/10.1016/j.cageo.2021.104708

Ayzel, G., Kurochkina, L., Abramov, D., & Zhuravlev, S. (2021). Development of a regional gridded runoff dataset using long short-term memory (LSTM) networks. *Hydrology, 8*, 6. https://doi.org/10.3390/hydrology8010006

Baudron, P., Alonso-Sarría, F., García-Aróstegui, J. L., Cánovas-García, F., Martínez-Vicente, D., & Moreno-Brotóns, J. (2013). Identifying the origin of groundwater samples in a multi-layer aquifer system with Random Forest classification. *Journal of Hydrology, 499*, 303–315. https://doi.org/10.1016/j.jhydrol.2013.07.009

Bergström, S. (1992). *The HBV model – its structure and applications* (Swedish Meteorological and Hydrological Institute, Norrköping 4) (pp. 1–33).

Beven, K. (1989). Changing ideas in hydrology — The case of physically-based models. *Journal of Hydrology, 105*, 157–172. https://doi.org/10.1016/0022-1694(89)90101-7

Bhasme, P., Vagadiya, J., & Bhatia, U. (2022). Enhancing predictive skills in physically-consistent-way: Physics informed machine learning for hydrological processes. *Journal of Hydrology, 615*, 128618. https://doi.org/10.1016/j.jhydrol.2022.128618

Biswal, B. (2016). Dynamic hydrologic modeling using the zero-parameter Budyko model with instantaneous dryness index. *Geophysical Research Letters, 43*, 9696–9703. https://doi.org/10.1002/2016GL070173

Biswal, B., & Marani, M. (2014). 'Universal' recession curves and their geomorphological interpretation. *Advances in Water Resources, 65*, 34–42. https://doi.org/10.1016/j.advwatres.2014.01.004

Biswal, B., & Nagesh Kumar, D. (2015). Estimation of 'drainable' storage—A geomorphological approach. *Advances in Water Resources, 77*, 37–43. https://doi.org/10.1016/j.advwatres.2014.12.009

Blöschl, G., & Sivapalan, M. (1995). Scale issues in hydrological modelling: A review. *Hydrological Processes, 9*, 251–290. https://doi.org/10.1002/hyp.3360090305

Booker, D. J., & Woods, R. A. (2014). Comparing and combining physically-based and empirically-based approaches for estimating the hydrology of ungauged catchments. *Journal of Hydrology, 508*, 227–239. https://doi.org/10.1016/j.jhydrol.2013.11.007

Boser, B. E., Guyon, I. M., & Vapnik, V. N. (1992). A training algorithm for optimal margin classifiers. In *Proceedings of the fifth annual workshop on computational learning theory* (pp. 144–152). ACM. https://doi.org/10.1145/130385.130401

Brutsaert, W., & Nieber, J. L. (1977). Regionalized drought flow hydrographs from a mature glaciated plateau. *Water Resources Research, 13*, 637–643. https://doi.org/10.1029/WR013i003p00637

Buch, A. M., Mazumdar, H. S., & Pandey, P. C. (1993). Application of artificial neural networks in hydrological modeling: A case study of runoff simulation of a Himalayan glacier basin. In

Proceedings of 1993 international conference on neural networks (IJCNN-93-Nagoya, Japan) (pp. 971–974). IEEE. https://doi.org/10.1109/IJCNN.1993.714073

Caraway, N. M., McCreight, J. L., & Rajagopalan, B. (2014). Multisite stochastic weather generation using cluster analysis and k-nearest neighbor time series resampling. *Journal of Hydrology, 508*, 197–213. https://doi.org/10.1016/j.jhydrol.2013.10.054

Ch, S., Anand, N., Panigrahi, B. K., & Mathur, S. (2013). Streamflow forecasting by SVM with quantum behaved particle swarm optimization. *Neurocomputing, 101*, 18–23. https://doi.org/10.1016/j.neucom.2012.07.017

Chiang, J. L., & Tsai, Y. S. (2011). Reservoir drought prediction using support vector machines. *Applied Mechanics and Materials, 145*, 455–459. https://doi.org/10.4028/www.scientific.net/AMM.145.455

Cho, K., & Kim, Y. (2022). Improving streamflow prediction in the WRF-hydro model with LSTM networks. *Journal of Hydrology, 605*, 127297. https://doi.org/10.1016/j.jhydrol.2021.127297

Daw, A., Karpatne, A., Watkins, W., Read, J., & Kumar, V. (2017). Physics-guided neural networks (PGNN): An application in lake temperature modeling. https://doi.org/10.48550/arXiv.1710.11431

Daw, A., Thomas, R. Q., Carey, C. C., Read, J. S., Appling, A. P., & Karpatne, A. (2020). Physics-guided architecture (PGA) of neural networks for quantifying uncertainty in Lake temperature modeling. In *Proceedings of the 2020 SIAM international conference on data mining* (pp. 532–540). Society for Industrial and Applied Mathematics. https://doi.org/10.1137/1.9781611976236.60

Dechter, R. (1986) Learning while searching in constraint satisfaction problems. In *AAAI-86 Proceedings* (pp. 179–183).

Ding, A., Zhang, Q., Zhou, X., & Dai, B. (2016). Automatic recognition of landslide based on CNN and texture change detection. In *2016 31st Youth Academic annual conference of Chinese Association of Automation (YAC)* (pp. 444–448). IEEE. https://doi.org/10.1109/YAC.2016.7804935

Feng, Z., Niu, W., Zhang, R., Wang, S., & Cheng, C. (2019). Operation rule derivation of hydropower reservoir by k-means clustering method and extreme learning machine based on particle swarm optimization. *Journal of Hydrology, 576*, 229–238. https://doi.org/10.1016/j.jhydrol.2019.06.045

Freeze, R. A., & Harlan, R. L. (1969). Blueprint for a physically-based, digitally-simulated hydrologic response model. *Journal of Hydrology, 9*, 237–258. https://doi.org/10.1016/0022-1694(69)90020-1

Friedl, M. A., & Brodley, C. E. (1997). Decision tree classification of land cover from remotely sensed data. *Remote Sensing of Environment, 61*, 399–409. https://doi.org/10.1016/S0034-4257(97)00049-7

Guo, J., Zhou, J., Qin, H., Zou, Q., & Li, Q. (2011). Monthly streamflow forecasting based on improved support vector machine model. *Expert Systems with Applications, 38*, 13073–13081. https://doi.org/10.1016/j.eswa.2011.04.114

Halff, A. H., Halff, H. M., & Azmoodeh, M. (1993). Predicting runoff from rainfall using neural networks. In *Proceedings engineering hydrolgy* (pp. 760–765). American Society of Civil Engineers.

Han, D., Kwong, T., & Li, S. (2007). Uncertainties in real-time flood forecasting with neural networks. *Hydrological Processes, 21*, 223–228. https://doi.org/10.1002/hyp.6184

Hochreiter, S., & Schmidhuber, J. (1997). Long short-term memory. *Neural Computation, 9*, 1735–1780. https://doi.org/10.1162/neco.1997.9.8.1735

Holman, D., Sridharan, M., Gowda, P., Porter, D., Marek, T., Howell, T., & Moorhead, J. (2014). Gaussian process models for reference ET estimation from alternative meteorological data sources. *Journal of Hydrology, 517*, 28–35. https://doi.org/10.1016/j.jhydrol.2014.05.001

Hsu, K., Gupta, H. V., & Sorooshian, S. (1995). Artificial neural network modeling of the rainfall-runoff process. *Water Resources Research, 31*, 2517–2530. https://doi.org/10.1029/95WR01955

Istalkar, P., Kadu, A., & Biswal, B. (2023). Value of process understanding in the era of machine learning: A case for recession flow prediction. *Journal of Hydrology, 130350*. https://doi.org/10.1016/j.jhydrol.2023.130350

Jehanzaib, M., Bilal Idrees, M., Kim, D., & Kim, T.-W. (2021). Comprehensive evaluation of machine learning techniques for hydrological drought forecasting. *Journal of Irrigation and Drainage Engineering, 147*. https://doi.org/10.1061/(ASCE)IR.1943-4774.0001575

Jia, X., Willard, J., Karpatne, A., Read, J. S., Zwart, J. A., Steinbach, M., & Kumar, V. (2021). Physics-guided machine learning for scientific discovery: An application in simulating Lake temperature profiles. *ACM/IMS Transactions on Data Science, 2*, 1–26. https://doi.org/10.1145/3447814

Kadu, A., & Biswal, B. (2022). A model combination approach for improving streamflow prediction. *Water Resources Management, 36*, 5945–5959. https://doi.org/10.1007/s11269-022-03336-5

Kaiser, E., Kutz, J. N., & Brunton, S. L. (2018). Sparse identification of nonlinear dynamics for model predictive control in the low-data limit. *Proceedings of the Royal Society A: Mathematical, Physical and Engineering Sciences, 474*, 20180335. https://doi.org/10.1098/rspa.2018.0335

Kampf, S. K., & Burges, S. J. (2007). Parameter estimation for a physics-based distributed hydrologic model using measured outflow fluxes and internal moisture states. *Water Resources Research, 43*. https://doi.org/10.1029/2006WR005605

Karpatne, A., Atluri, G., Faghmous, J. H., Steinbach, M., Banerjee, A., Ganguly, A., Shekhar, S., Samatova, N., & Kumar, V. (2017). Theory-guided data science: A new paradigm for scientific discovery from data. *IEEE Transactions on Knowledge and Data Engineering, 29*, 2318–2331. https://doi.org/10.1109/TKDE.2017.2720168

Khandelwal, A., Xu, S., Li, X., Jia, X., Stienbach, M., Duffy, C., Nieber, J., & Kumar, V. (2020) Physics guided machine learning methods for hydrology.

Kim, D. Y., & Song, C. M. (2020). Developing a discharge estimation model for ungauged watershed using CNN and hydrological image. *Water, 12*, 3534. https://doi.org/10.3390/w12123534

Kolachian, R., & Saghafian, B. (2021). Hydrological drought class early warning using support vector machines and rough sets. *Environment and Earth Science, 80*, 390. https://doi.org/10.1007/s12665-021-09536-3

Kratzert, F., Klotz, D., Brenner, C., Schulz, K., & Herrnegger, M. (2018). Rainfall–runoff modelling using long short-term memory (LSTM) networks. *Hydrology and Earth System Sciences, 22*, 6005–6022. https://doi.org/10.5194/hess-22-6005-2018

Kratzert, F., Klotz, D., Herrnegger, M., Sampson, A. K., Hochreiter, S., & Nearing, G. S. (2019a). Toward improved predictions in ungauged basins: Exploiting the power of machine learning. *Water Resources Research, 55*, 11344–11354. https://doi.org/10.1029/2019WR026065

Kratzert, F., Klotz, D., Shalev, G., Klambauer, G., Hochreiter, S., & Nearing, G. (2019b). Towards learning universal, regional, and local hydrological behaviors via machine learning applied to large-sample datasets. *Hydrology and Earth System Sciences, 23*, 5089–5110. https://doi.org/10.5194/hess-23-5089-2019

Lecun, Y., Bottou, L., Bengio, Y., & Haffner, P. (1998). Gradient-based learning applied to document recognition. *Proceedings of the IEEE, 86*, 2278–2324. https://doi.org/10.1109/5.726791

Lee, J.-Y., & Kim, J.-S. (2021). Detecting areas vulnerable to flooding using hydrological-topographic factors and logistic regression. *Applied Sciences, 11*, 5652. https://doi.org/10.3390/app11125652

Li, X., & Willems, P. (2019). Probabilistic flood prediction for urban sub-catchments using sewer models combined with logistic regression models. *Urban Water Journal, 16*, 687–697. https://doi.org/10.1080/1573062X.2020.1726409

Li, W., Kiaghadi, A., & Dawson, C. (2021). High temporal resolution rainfall–runoff modeling using long-short-term-memory (LSTM) networks. *Neural Computing and Applications, 33*, 1261–1278. https://doi.org/10.1007/s00521-020-05010-6

Lin, L., & Weng, F. (2018). Estimation of hurricane maximum wind speed using temperature anomaly derived from advanced technology microwave sounder. *IEEE Geoscience and Remote Sensing Letters, 15*, 639–643. https://doi.org/10.1109/LGRS.2018.2807763

Mair, A., & El-Kadi, A. I. (2013). Logistic regression modeling to assess groundwater vulnerability to contamination in Hawaii, USA. *The Journal of Contaminant Hydrology, 153*, 1–23. https://doi.org/10.1016/j.jconhyd.2013.07.004

Moore, R. J. (2007). The PDM rainfall-runoff model. *Hydrology and Earth System Sciences, 11*, 483–499. https://doi.org/10.5194/hess-11-483-2007

Nayak, A. K., Biswal, B., & Sudheer, K. P. (2022). Drought hotspot maps and regional drought characteristics curves: Development of a novel framework and its application to an Indian River basin undergoing climatic changes. *Science of the Total Environment, 807*, 151083. https://doi.org/10.1016/j.scitotenv.2021.151083

Nearing, G. S., Kratzert, F., Sampson, A. K., Pelissier, C. S., Klotz, D., Frame, J. M., Prieto, C., & Gupta, H. V. (2021). What role does hydrological science play in the age of machine learning? *Water Resources Research, 57*. https://doi.org/10.1029/2020WR028091

Ozdemir, A. (2011). Using a binary logistic regression method and GIS for evaluating and mapping the groundwater spring potential in the Sultan Mountains (Aksehir, Turkey). *Journal of Hydrology, 405*, 123–136. https://doi.org/10.1016/j.jhydrol.2011.05.015

Pal, M., & Mather, P. M. (2003). An assessment of the effectiveness of decision tree methods for land cover classification. *Remote Sensing of Environment, 86*, 554–565. https://doi.org/10.1016/S0034-4257(03)00132-9

Patnaik, S., Biswal, B., Nagesh Kumar, D., & Sivakumar, B. (2015). Effect of catchment characteristics on the relationship between past discharge and the power law recession coefficient. *Journal of Hydrology, 528*, 321–328. https://doi.org/10.1016/j.jhydrol.2015.06.032

Pawar, S., San, O., Aksoylu, B., Rasheed, A., & Kvamsdal, T. (2021). Physics guided machine learning using simplified theories. *Physics of Fluids, 33*, 011701. https://doi.org/10.1063/5.0038929

Peel, M. C., & McMahon, T. A. (2020). Historical development of rainfall-runoff modeling. *WIREs Water, 7*. https://doi.org/10.1002/wat2.1471

Perrin, C., Michel, C., & Andréassian, V. (2003). Improvement of a parsimonious model for streamflow simulation. *Journal of Hydrology, 279*, 275–289. https://doi.org/10.1016/S0022-1694(03)00225-7

Prairie, J. R., Rajagopalan, B., Fulp, T. J., & Zagona, E. A. (2006). Modified K-NN model for stochastic streamflow simulation. *Journal of Hydrologic Engineering, 11*, 371–378. https://doi.org/10.1061/(ASCE)1084-0699(2006)11:4(371)

Ragettli, S., Zhou, J., Wang, H., Liu, C., & Guo, L. (2017). Modeling flash floods in ungauged mountain catchments of China: A decision tree learning approach for parameter regionalization. *Journal of Hydrology, 555*, 330–346. https://doi.org/10.1016/j.jhydrol.2017.10.031

Raghavendra, N. S., & Deka, P. C. (2014). Support vector machine applications in the field of hydrology: A review. *Applied Soft Computing, 19*, 372–386. https://doi.org/10.1016/j.asoc.2014.02.002

Rahimzad, M., Moghaddam Nia, A., Zolfonoon, H., Soltani, J., Danandeh Mehr, A., & Kwon, H.-H. (2021). Performance comparison of an LSTM-based deep learning model versus conventional machine learning algorithms for streamflow forecasting. *Water Resources Management, 35*, 4167–4187. https://doi.org/10.1007/s11269-021-02937-w

Random Forests. (2001). Random forest. *Machine Learning, 45*(1), 5–32.

Reddyvaraprasad, C., Patnaik, S., & Biswal, B. (2020). Recession flow prediction in gauged and ungauged basins by just considering past discharge information. *Hydrological Sciences Journal, 65*, 21–32. https://doi.org/10.1080/02626667.2019.1643465

Rosenblatt, F. (1958). The perceptron: A probabilistic model for information storage and organization in the brain. *Psychological Review, 65*, 386–408. https://doi.org/10.1037/h0042519

Rumelhart, D. E., Hinton, G. E., & Williams, R. J. (1986). Learning representations by back-propagating errors. *Nature, 323*, 533–536. https://doi.org/10.1038/323533a0

Samaniego, L., Bardossy, A., & Schulz, K. (2008). Supervised classification of remotely sensed imagery using a modified k-NN technique. *IEEE Transactions on Geoscience and Remote Sensing, 46*, 2112–2125. https://doi.org/10.1109/TGRS.2008.916629

Schmidhuber, J. (2015). Deep learning in neural networks: An overview. *Neural Networks, 61*, 85–117. https://doi.org/10.1016/j.neunet.2014.09.003

Schoppa, L., Disse, M., & Bachmair, S. (2020). Evaluating the performance of random forest for large-scale flood discharge simulation. *Journal of Hydrology, 590*, 125531. https://doi.org/10.1016/j.jhydrol.2020.125531

Smith, J., & Eli, R. N. (1995). Neural-network models of rainfall-runoff process. *Journal of Water Resources Planning and Management, 121*, 499–508. https://doi.org/10.1061/(ASCE)0733-9496(1995)121:6(499)

Song, Y. H., Chung, E.-S., & Shahid, S. (2022). Differences in extremes and uncertainties in future runoff simulations using SWAT and LSTM for SSP scenarios. *Science of the Total Environment, 838*, 156162. https://doi.org/10.1016/j.scitotenv.2022.156162

Sun, A. Y., Wang, D., & Xu, X. (2014). Monthly streamflow forecasting using Gaussian process regression. *Journal of Hydrology, 511*, 72–81. https://doi.org/10.1016/j.jhydrol.2014.01.023

Suryanarayana, C., Sudheer, C., Mahammood, V., & Panigrahi, B. K. (2014). An integrated wavelet-support vector machine for groundwater level prediction in Visakhapatnam, India. *Neurocomputing, 145*, 324–335. https://doi.org/10.1016/j.neucom.2014.05.026

Tang, Y., Zang, C., Wei, Y., & Jiang, M. (2019). Data-driven modeling of groundwater level with least-square support vector machine and spatial–temporal analysis. *Geotechnical and Geological Engineering, 37*, 1661–1670. https://doi.org/10.1007/s10706-018-0713-6

Todini, E. (2011). History and perspectives of hydrological catchment modelling. *Hydrology Research, 42*, 73–85. https://doi.org/10.2166/nh.2011.096

Turing, A. M. (1950). I.—Computing machinery and intelligence. *Mind, LIX*, 433–460. https://doi.org/10.1093/mind/LIX.236.433

Tyralis, H., Papacharalampous, G., & Langousis, A. (2019). A brief review of random forests for water scientists and practitioners and their recent history in water resources. *Water, 11*, 910. https://doi.org/10.3390/w11050910

Unnithan, S. L. K., Biswal, B., Rüdiger, C., & Dubey, A. K. (2024). A novel conceptual flood inundation model for large scale data-scarce regions. *Environmental Modelling & Software, 171*, 105863. https://doi.org/10.1016/j.envsoft.2023.105863

Vapnik, V. N. (2000). *The nature of statistical learning theory*. Springer. https://doi.org/10.1007/978-1-4757-3264-1

Xiang, Z., Yan, J., & Demir, I. (2020). A rainfall-runoff model with LSTM-based sequence-to-sequence learning. *Water Resources Research, 56*. https://doi.org/10.1029/2019WR025326

Xie, K., Liu, P., Zhang, J., Han, D., Wang, G., & Shen, C. (2021). Physics-guided deep learning for rainfall-runoff modeling by considering extreme events and monotonic relationships. *Journal of Hydrology, 603*, 127043. https://doi.org/10.1016/j.jhydrol.2021.127043

Xu, H., Ma, C., Lian, J., Xu, K., & Chaima, E. (2018). Urban flooding risk assessment based on an integrated k-means cluster algorithm and improved entropy weight method in the region of Haikou, China. *Journal of Hydrology, 563*, 975–986. https://doi.org/10.1016/j.jhydrol.2018.06.060

Yang, J., Jakeman, A., Fang, G., & Chen, X. (2018). Uncertainty analysis of a semi-distributed hydrologic model based on a Gaussian process emulator. *Environmental Modelling & Software, 101*, 289–300. https://doi.org/10.1016/j.envsoft.2017.11.037

Zhang, Y., Ragettli, S., Molnar, P., Fink, O., & Peleg, N. (2022). Generalization of an encoder-decoder LSTM model for flood prediction in ungauged catchments. *Journal of Hydrology, 614*, 128577. https://doi.org/10.1016/j.jhydrol.2022.128577

A Review of Approaches and Applications for Streamflow Forecasting Using AI-Based Models

Manish K. Nema and G. E. Nagashree

Abstract Streamflow forecasting is one of the most crucial exercises for reservoir operations, hydropower generation, and irrigation scheduling management. Conventionally, it is done by developing a forecasting model based on historical records of various variables. Broadly, the models can be divided into either physically based mathematical models or data-driven "black box" models. To understand the hydrologic system, the development of physically based models necessitates a vast quantity of diverse data connected to numerous physical processes, which is typically challenging to gather under actual field conditions, especially in developing nations. On the other hand, data-driven models do not require system expertise, and as a result, they have gained popularity, particularly with improvements in artificial intelligence technology. Stochastic time series models have historically been employed to predict streamflow. Because Artificial Neural Network (ANN) models forecast streamflow more accurately than time series models did in the 1990s, they became increasingly popular. Artificial intelligence (AI), an offshoot of computer science, can analyse long-series and large-scale hydrological data. In recent years, applying AI technology to hydrological forecasting modelling has been one of the front-burner issues. Streamflow forecasting also uses various AI techniques like SVM, Fuzzy Logic, GP, GEP, etc. Later, by merging them with other models, the performance of the current AI-based data-driven models is improved. This chapter discusses various AI-based models, their basic concepts, types, and applications for forecasting streamflow.

M. K. Nema (✉)
National Institute of Hydrology, Roorkee, India

G. E. Nagashree
Indian Institute of Technology, Roorkee, India

R. Srivastav, P. C. Nayak (eds.), *Applications of Machine Learning in Hydroclimatology*, The Springer Series in Applied Machine Learning,
https://doi.org/10.1007/978-3-031-64403-0_2

1 Introduction

Today, the world is facing a significant challenge in managing its water resources. The ever-increasing human population has caused fierce competition for water consumption globally. With only 3% of the world's water being portable fresh water, it is essential to have an effective management system in place to ensure the sustainable use of this precious resource. Water resource management can be defined as the process of preparing, creating, supplying, and controlling the optimal use of water resources. Dams play a critical role in preventing natural water-related disasters and managing the amount of freshwater stored in their associated reservoirs, which are crucial for water resource management. Reservoirs are the most efficient water storage facilities, and their effective management is necessary for the natural distribution of water, both spatially and temporally. In addition to the rainfall that provides water for hydroelectricity generation and irrigation, it is vital to mitigate strong inflows to prevent floods and to devise drought mitigation plans in case of minimal precipitation. However, forecasting streamflow is challenging due to the highly variable, non-stationary, dynamic, and non-linear behaviour of water flow. Effective reservoir management requires accurate predictions of the potential reservoir inflow, which can be achieved through streamflow forecasting.

Streamflow forecasting is divided into two categories: long-term forecasting and short-term forecasting. Long-term forecasting predicts water flow on a weekly, monthly, or yearly basis and is essential for planning long-term hydro-power generation and water releases for drought mitigation. On the other hand, short-term forecasting predicts water flow hourly or daily.

The technological applications in reservoir forecasting have been increasing over the past few decades, with the introduction of computers and other technological devices in the late 1970s. During this time, conventional time series models, such as the autoregressive and moving average (ARMA) model, were widely used for reservoir forecasting. ARMA reveals the causal relationship between events that occurred in the past and events that are yet to happen. In addition to the ARMA model, other models such as linear regression (LR), multiple linear regression (MLR), and polynomial linear regression (PLR) were also used in reservoir forecasting. However, these conventional models failed to capture the non-linearity in the recorded data, resulting in low forecasting accuracy that significantly impacts the short-term and long-term planning for reservoirs.

This issue of low accuracy in forecasting led to the development of improved and more accurate models, especially with the advent of the digital revolution and the growing use of artificial intelligence (AI) in hydrology and water management. AI models can make accurate predictions with little input or even without data. This attribute of the AI-based model makes it a valuable tool in hydrology and has been used in several methods for predicting hydrological parameters.

Table 1 gives information on the application of various AI models in streamflow forecasting and the significance of these studies will be discussed later. This chapter highlights AI and hybrid models, and their application and outcomes are noted.

Table 1 Information of previous studies on AI models for streamflow forecasting

Author	Model/Algorithm	Input variable	Evaluation criteria	Time scale
Imrie et al. (2000)	BP and CC learning algorithm	Streamflow	R2,	Daily
Toth et al. (2000)	ARMA, KNN, ANN	Rainfall	Efficiency	Hourly
Valença and Ludermir (2001)	Comparison between the numerical model (PARMA) and NN model (NSRBN)	Inflow	AAPE	Weekly
Dolling and Varas (2002)	Used different learning algorithms	ENSO index, temperature, precipitation, snow course		Monthly
Birikundavyi et al. (2002)	MLP, PERVIS, ARMAX, Kalman filter	Streamflow, rainfall, temperature	Nash, RMSE	Daily
Cigizoglu (2005)	GRNN, FFBP	Daily mean flow series	MSE	Daily
Kişi (2007)	Backpropagation, cascade correlation, conjugate gradient, Levenberg-Marquardt	Inflow with three antecedent values	RMSE, R	Daily
Firat and Mahmud (2007)	ANFIS	Streamflow	Relative error, E, R, RMSE	Daily
Anctil et al. (2008)	MLP	Rainfall, soil moisture, SMI streamflow	MAE, RMSE, EI, PI, R2	Daily
Makkeasorn et al. (2008)	GP, ANN	SST, rainfall (NEXARD), meteorological data, streamflow	RMSE, R2, PE (mean percentage error)	Daily
El-Shafie et al. (2009)	RFBNN, MLP, ARMA	Streamflow	MSE, FE, RE, PCF, LCF	Monthly
Pramanik and Panda (2009)	ANN (CG, LM, GD) and ANFIS	Streamflow	R, R^2, RMSE, modelling efficiency, percentage deviation in peak flow	Daily
Akhtar et al. (2009)	ANN	Pre-processed rainfall data (satellite), streamflow, ET	RMSE, PRSE, MAE	Daily
Wang et al. (2009)	ANN, ARMA, GP, ANFIS, SVM	Streamflow	R, Nash, RMSE, MAPE	Monthly
Besaw et al. (2010)	GRNN, CPN, MLR, and ARMAX	Rainfall, temperature, streamflow	R2, Nash, RMSE	Daily
Kashid et al. (2010)	Genetic programming, ANN, LR	From ENSO, EQUINOO indices, lag	R2, Nash	Weekly

(continued)

Table 1 (continued)

Author	Model/Algorithm	Input variable	Evaluation criteria	Time scale
		rainfall and OLR-rainfall are predicted, and then this data is used to predict streamflow		
Abudu et al. (2011)	TFN, ANN, TFN + ANN	Snow telemetry, SWE, rainfall	NRMSE, R2, RMSE	Monthly
Ish et al. (2012)	GP, SVR	ENSO, EQUINOO indices, lag rainfall and OLR-	RMSE, R2	Weekly
Danandeh et al. (2013)	WANN, LGP	Upstream, downstream flow	RMSE, NSE	Monthly
Danandeh et al. (2014)	LGP, FFBP, GRNN, RBF	Upstream, downstream flow	RMSE, NSE	Monthly
Augusto et al. (2014)	DWT-ANN, ANN	Streamflow	RMSE, MARE, CE	Daily
Huang et al. (2014)	M-EMDSVM, EMD-SVM, ANN, SVM	Streamflow	RMSE, MAPE, MAE	Monthly
Liu et al. (2014)	DWT-SVR, SVR, ensemble	Streamflow	RMSE, NSE	Daily, monthly
Zamani et al. (2017)	ANN, ARIMA, ANN-ARIMA, ANFIS	Snowpack data, precipitation index, streamflow	R, RMSE	Monthly
Mundher et al. (2017)	ANFIS-FFA, ANFIS	Streamflow	MAE, RMSE, R, normalised goodness-of-fit	Monthly
Ravansalar et al. (2017)	WLGP, LGP, WANN, ANN, MLR	Streamflow	RMSE, MAE	Monthly
Wang and Qiu (2018)	EEMD-ARIMA	Streamflow	RMSE, MAE, MAPE, R	Monthly
Muhammad et al. (2019b)	OP-ELM, M5Tree, MARS, ANFSI-PSO	Precipitation and streamflow	RMSE, MAE, NSE, CA	Daily
Karenina et al. (2019)	DWT-ANN	Streamflow	R2, NSE, RMSE	Daily
Zuo et al. (2020)	SF-VMD-LSTM, SFMIS			
Chu et al. (2021)	FCM-PMI-DBN framework	Precipitation	RMSE, MAE, made, R2	Daily
Shu et al. (2021)	CNN, ELM, ANN	Streamflow, rainfall, atmospheric circulation factor	NSE, RMSE, MAE	Monthly
Kilinc and Haznedar (2022)	GA-LSTM	Streamflow	RMSE, MAE, MAPE, R2	Daily

Section 2 covers different streamflow forecasting techniques; Sect. 3 explains the concept of various AI and hybrid models with a few case studies; and Sect. 4 presents significant findings from previous studies. Incorporating these points may help in future modelling, and this chapter ends with conclusions in Sect. 5.

2 Streamflow Forecasting Techniques

Streamflow forecasting is predicting the rate of water flowing in a stream or river. It is an essential aspect of water resource management. Precise predictions allow for the optimal use of water systems, factoring in technical, financial, legal, and political considerations. A forecasting system that considers all relevant temporal and spatial variations in streamflow is vital for effectively controlling and managing water resources. Based on time, streamflow forecasting can be classified into short-term and long-term forecasts. Short-term forecasts, usually with lead times of hours and days, are commonly used for flood warnings and real-time water system operations. Long-term forecasts, with lead times ranging from weeks to months, are utilised for water system planning for hydro-power and addressing and mitigating droughts.

There are various methods used for streamflow forecasting, which can be categorised into (a) statistical method, (b) physical-based models, (c) conceptual models, and (d) artificial intelligence (AI)-based models, as shown in Fig. 1.

Statistical methods such as Time series analysis, Regression analysis, the Autoregressive (AR) model such as Auto-regressive Moving Average (ARMA) model, Auto-regressive Integrated Moving Average (ARIMA) model, ARMA with

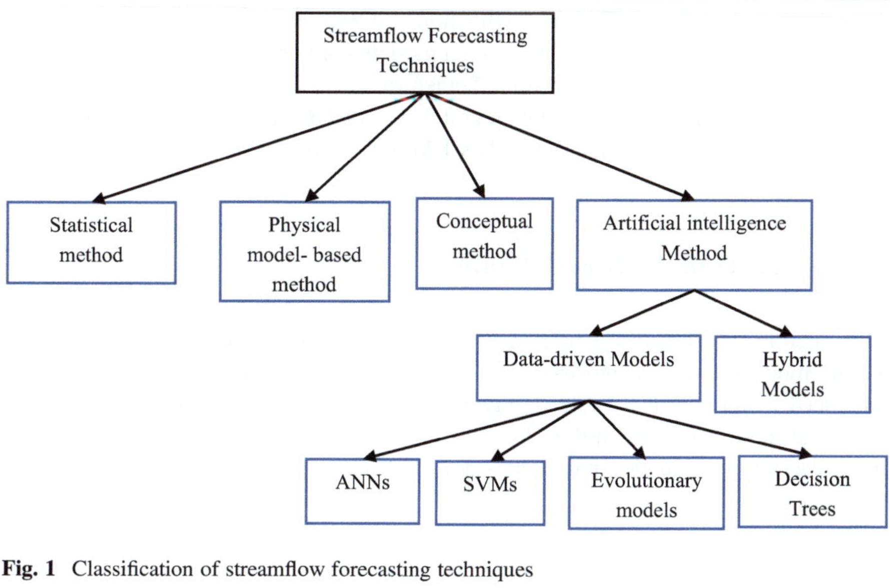

Fig. 1 Classification of streamflow forecasting techniques

exogeneous variables (ARMAX), and Bayesian modelling have been widely used for streamflow forecasting. These models use historical streamflow data and meteorological inputs to make predictions.

Physical-based models, such as the Soil and Water Assessment Tool (SWAT) model, Hydrological Engineering Centre's Hydrologic Modelling System (HCE-HMS), Hydrological Simulation Program-FORTRAN (HSPF), Distributed Hydrological Soil Vegetation Model (DHSVM), Catchment Land Surface Model (CLSM), Hydrological Predictive uncertainty Model (Hydro-PUM), use the fundamental physical processes of the hydrologic cycle to stimulate water movement through the watershed. To make predictions, these models require detailed information on the watershed, such as land use, soil type, vegetation, and meteorological data.

Conceptual models, such as the Muskingum method, Soil Conservation Service Curve Number (SCS-CN) method, and Hydrologiska Byråns Vattenbalansavdelning (HBV) model, use simple mathematical equations to describe the flow of water in a river. These models are based on the concept of storage and routing and require information on the river's geometry and flow characteristics. These models have some disadvantages, like limited ability to account for changes in the system, complex interactions between various hydrological processes, non-linear processes and feedback, and the need for expert knowledge and skill to use and interpret the model correctly. AI models are employed for streamflow forecasting to overcome these limitations.

AI-based models have gained popularity recently due to their potential to improve streamflow forecasting accuracy, making valuable water resource management tools. These models can be labelled as data-driven models and hybrid models. Data-driven models include the different forms of Artificial Neural Networks (ANNs), Support Vector Machines (SVM), Evolutionary models and Decision trees like Random Forest (RF), Classification and Regression Trees (CART). Hybrid models are the combination of data-driven models with or without the application of pre-processing methods like Wavelet Transforms (WT) and Empirical Decomposition Methods (EMD) like Artificial Neural Network-Genetic Algorithm (ANN-GA), ANN-SVM, ANN-RF, ANN-WT, ANN-EMD, and ANN-PSO.

3 Concepts of AI-Based Models

This section will discuss a brief and basic concept of various AI-based models. These models were categorised into data-driven and hybrid models based on advanced computation. Table 1 gives a brief overview of the application of AI-based models for streamflow forecasting, and it includes the choice of models, input variables, evaluation parameters and time-scale considerations.

3.1 Data-Driven Models

These models use historical streamflow data and other meteorological data to develop models that can predict future streamflow. These models used techniques such as statistical regression, machine learning, and artificial intelligence to find patterns and relationships in the data that can be used to make forecasts. Based on the working principle, data-driven models can be further classified into Neural Networks (NNs), SVMs, Decision trees, and Evolutionary Algorithms.

3.1.1 Artificial Neural Networks (ANNs)

ANNs are the most commonly used machine learning models inspired by the human brain's structure and function. They consist of interconnected layers of artificial neurons organised into input, hidden, and output layers. The input layer receives the input data, the hidden layer processes the data, and the output layer produces the final output. The neurons in the hidden layers use non-linear activation functions, for example, *sigmoidal, tanh,* and rectified linear units (ReLU), to process the input data, which allows ANNs to model complex relationships between inputs and outputs. The ANN's performance is determined by a set of weights and biases assigned to the network's neurons. These weights and biases are adjusted during training to minimise errors between the predicted and actual output.

The training method defines the purpose of the network. The most used training method is backpropagation (BP), a form of gradient descent method that propagates the loss as a signal into the network and updates network weights based on the chain rule. The other training algorithm is conjugate gradient (CG), which is an optimisation algorithm that is a faster and more efficient alternative to backpropagation but can be challenging to implement. *Levenberg-Marquardt (LM)*—which combines the gradient descent and Gauss-Newton methods to train ANNs and it converges faster, *Scaled Conjugate Gradient (SCG)* is a variation of the conjugate gradient algorithm that uses a scaling factor to adjust the learning rate of the network and can be used to train feedforward and recurrent neural networks, *Resilient backpropagation (RBP)*—uses a robust learning algorithm that can withstand the presence of noise in the input, and *Rprop algorithm*—is a variation of the resilient backpropagation algorithm that uses the sign of the gradient to adjust the learning rate, rather than the magnitude.

Based on the depth/number of hidden layers, ANNs can be subdivided into Shallow and Deep Neural Networks, the former consists of one hidden layer and the latter consists of more than one hidden layer. These neural networks are basically feedforward, i.e., they consist of layers of interconnected nodes or neurons where information flows through the network in only one direction, from input to output, without loops. Feedforward neural networks (FNN) used for streamflow forecasting are multi-layer perceptron (MLP), Radial Basis function feedforward neural network

(RBFNN), Recurrent Neural Network (RNN), Long-short Term Memory (LSTM), Gated Recurrent Unit (GRU), and Convolutional Neural Network (CNN).

MLP is a type of FNN composed of multiple layers of artificial neurons, like the essential ANN. RBFNN uses radial basis functions (RBFs) as activation functions which are placed at the input layer and trained using backpropagation. These RBFs are typically utilised for non-linear function approximation and classification tasks. RNN is designed to process sequential data. Unlike traditional FNNs, RNNs can maintain a hidden state that is passed from one time step to the next. This allows the network to process data effectively with a temporal or sequential aspect. RNNs can be further classified into different types, such as LSTM and GRU, which have been developed to address a specific issue in processing sequential data. LSTM is capable of learning and remembering long-term dependencies in sequential data. They can overcome the vanishing gradient problem that occurs in traditional RNNs by using gating mechanisms that selectively allow information to flow through the network. Like LSTM, GRUs are designed to handle the problem of vanishing gradients in traditional RNNs by introducing gates. GRUs have two gates; the update gate-determines how much of the previous state should be retained, while the reset gate determines how much should be forgotten. CNN is designed to process data with a grid-like topology, such as an image. The main building blocks of a CNN are convolutional layers, pooling layers, and fully connected layers. The convolutional layer applies a convolution operation to the input data, which allows the network to learn the local pattern in the data. The pooling layers are used to reduce the spatial dimensions of the data, which helps to reduce the computational cost and improve the generalisation of the network. The fully connected layers are used to make the final prediction based on the features learned by the convolution and pooling layers.

Case Studies

Intercomparison of accuracies of short-term rainfall forecasts was done by (Toth et al., 2000) using the outputs from ARMA models, ANNs, and the nearest-neighbours method. ANNs show significant improvement in flood forecasting accuracy compared to heuristic approaches in a case study in Italy.

A method for improving the generalisation ability of ANN in river flow prediction by adding a guidance system to cascade-correlation learning (CC) algorithm architecture was proposed by (Imrie et al., 2000). The modified algorithm produced better ANN models than the standard BP algorithm and the original CC algorithm. The study highlights the importance of an appropriate output activation function for extrapolating beyond the calibration data.

In a streamflow forecasting study, Valença and Ludermir (2001) show the comparison between two models for river flow forecasting, the Non-linear Sigmoidal Regression Blocks Networks (NSRBN) and the Periodic Auto Regressive Moving Average (PRMA) models. The result shows that NSRBN reduces forecasting error by at least 19% and provides more accurate predictions than PARMA models. NSRBN algorithms use adaptive synthesis to solve the subjective choice of network architecture.

The performance of neural networks in forecasting daily streamflow with a conceptual model and an autoregressive model with a Kalman filter was compared by Birikundavyi et al. (2002). Results show that NNs outperform the other models by up to 5-day-ahead forecasts based on mean square errors and the Nash coefficient. Also recommends that further model improvement can be achieved by eliminating weights and bias, using RBF networks, and ensuring reliable input data.

The performance of generalised regression neural network (GRNN) and FFBP methods were compared in forecasting and estimating intermittent daily mean flows (Cigizoglu, 2005). The results showed that GRNN was superior to FFBP, regression, and stochastic methods regarding selected performance criteria. GRNN did not face the local minima problem faced by FFBP and produced positive and physically plausible results, while FFBP provided negative approximations for low flows.

In another study, the performance of four different ANN algorithms in forecasting daily streamflow was investigated (Kişi, 2007). The result showed that the LM algorithm took less time for training than the other three algorithms, while the BP algorithm took an unusually high number of iterations and time for training.

3.1.2 Support Vector Machine (SVM)

SVM is based on the concept of finding a hyperplane that maximally separates different classes of data points in a high-dimensional feature space. It constructs a boundary that is as possible from the closest data points of different classes, which are called support vectors. The key advantage of SVM is that it can handle high-dimensional and non-linearly separable data using *Kernels*, which transform the data into a higher-dimensional space where it may become linearly separable. The kernel trick is used to avoid the computational cost of calculating the data coordinates in high-dimensional space. Several types of kernels can be used in SVM, including *linear, polynomial, and radial basis function* (RBF) kernels. The choice of kernel depends on the specific problem and type of data. SVM can also be used for anomaly detection and outlier detection by setting the threshold of the decision function to a lower value and considering the points with a higher decision function value as an outlier. SVM can be used for classification and regression tasks.

Case Studies

The study by Ish et al. (2012) describes the use of two AI approaches, SVR and GP, to model the complex relationships between large-scale atmospheric circulation patterns and the streamflow of the Mahanadi River basin. The SVR model was used to forecast the monthly basin-scale streamflow and showed better results than traditional statistical models (ARIMA and ARMA). SVR used only single variable input and has the potential to yield better results.

3.1.3 Decision Trees

A decision tree is a flowchart-like tree structure where an internal node represents a feature (or attribute), the branch represents a decision rule, and each leaf node represents the outcome. The topmost node in the decision tree is known as the root node. It learns to partition the data into subsets based on the values of the input features. Each internal node of the decision tree corresponds to a test on an attribute, each branch corresponds to the outcome of a test, and each leaf node represents a class label (decision taken after computing all attributes). The goal of a decision tree is to create a training model which can be used to predict the class or value of the target variable by learning decision rules inferred from training data. Random Forest (RF), Gradient boosting, Extreme Gradient Boosting (XGB), classification and regression trees (CART), M5.

3.1.4 Fuzzy Logic

Fuzzy logic is a mathematical method for representing and manipulating uncertain or imprecise information. It is based on the idea of fuzzy sets, in which an element can belong to a set to a certain degree rather than just entirely in or out of the set, as in traditional binary logic. The main components of the fuzzy logic model are 1) fuzzy rules—these are if-then rules that express the relationship between input and output variables using fuzzy sets; 2) fuzzification—the process of converting crisp input values into fuzzy sets, 3) inference engine—the process of applying fuzzy rules to the input sets to produce fuzzy output sets; 4) defuzzification—the process of converting fuzzy output sets into crisp values, and 5) Membership functions—are mathematical functions that define the membership of a value to a fuzzy set.

Mamdani and Sugeno's methods are two different approaches to fuzzy logic control. The Mamdani method is based on the use of fuzzy sets and membership functions to represent linguistic variables and their relationships. It involves using fuzzy if-then rules to infer a conclusion from a group of input variables. On the other hand, the Sugeno method is based on using fuzzy numbers and fuzzy arithmetic to represent the same information. It involves using a set of fuzzy if-then rules to generate a numerical output. Fuzzy logic can be combined with other models ANN to get ANFIS which improves the accuracy and robustness of streamflow forecasts.

Case Studies

The ANFIS approach for river flow forecasting (Firat & Mahmud, 2007) was investigated, wherein different input structures were tested, and the best structure was selected. Four different training/testing data sets were also constructed and evaluated. The results showed that ANFIS is an effective method for river flow estimation, providing high accuracy and reliability.

In another study (Pramanik & Panda, 2009), the ANFIS model produced a higher correlation coefficient, lower RMSE, and higher modelling efficiency than the best ANN model. In conclusion, the study suggests that the ANFIS technique is a better choice for barrage outflow estimation than ANN in the Mahanadi River basin case and that considering reservoir release data from antecedent time steps can further improve the accuracy of the prediction.

3.1.5 Evolutionary Algorithms (EA)

Evolutionary algorithms (EAs) are a family of optimisation algorithms inspired by natural evolution principles. They are used to find approximate solutions to optimisation and search problems. The basic idea behind evolutionary algorithms is to stimulate the process of natural selection to find the best solution to a given problem. The critical components of EAs include population, fitness function, selection, crossover, mutation, replacement, and termination. There are several different types of evolutionary algorithms, such as genetic algorithms (GA), linear genetic programming (LGP), and particle swarm optimisation (PSO).

Case Studies

The results from a study (Makkeasorn et al., 2008) showed that the NN models performed well in predicting streamflow with sea surface temperature (SST) and meteorological data, but GP models outperformed them. The pre-selected input variables improved the performance of NN models but were generally worse than GP-derived models.

Another study used a GP model to predict weekly gridded monsoon rainfall in the Mahanadi catchment in India (Kashid et al., 2010) by considering large-scale atmospheric patterns, lagged rainfall information, and basin-scale OLR anomaly. The GP-predicted rainfall was then used as input for the GP-derived rainfall-runoff models to improve the weekly streamflow forecast compared to models using only observed rainfall.

The study by Danandeh et al. (2013) aimed to compare the performance of hybrid wavelet-ANN (WANN) and linear genetic programming (LGP) in monthly streamflow prediction. The results showed that LGP had higher performance than WANN in all reference models and pointed out that the number of inputs for WANN models can affect the accuracy of the output, and the optimisation of the input signals is easier with LGP.

The SVM-based study (Sudheer et al., 2014) aimed to predict monthly streamflow and particle swarm optimisation (PSO) techniques for tuning SVM parameters. The study found that the choice of input structure was essential for SVM performance, and various statistical performances were used to determine the best input structure. Overall, the results suggested that SVMs can be a promising alternative for forecasting monthly streamflow, especially when dealing with non-linear data patterns.

3.2 *Hybrid Models*

Hybrid models combine two or more AI technologies to take advantage of the strengths of each technique. It may connect a physical model and statistical model, neural networks with an evolutionary algorithm, fuzzy logic, and decision tree to improve the forecast accuracy by incorporating both models' strengths.

Hybrid models also include advanced pre-processing techniques like wavelet transformation (WT) and empirical mode decomposition (EMD) to reduce the noise in the data so that the model produces accurate results. WT is a mathematical method that uses wavelets to represent signals in a compact form. It is well suited for handling signals with well-defined frequency content. It provides a multi-resolution analysis of the data and can be helpful in denoising data with Gaussian or Poisson noise. The main types of WT used for streamflow forecasting are, continuous wavelet transform (CWT)—which operates on constant signals, and discrete wavelet transform (DWT)—which works on discrete signals (Augusto et al., 2014).

EMD is a data-driven approach that can adapt the data's intrinsic mode functions (IMF), leading to better denoising results compared to WT. IMFs are characterised by having both zero mean and similar energy in the upper and lower envelopes. Each IMF represents a different signal characteristic, such as its high-frequency, low-frequency, or time-varying components. By applying the EMD algorithm, the signal is separated into multiple IMFs, which can be used for feature extraction or noise reduction. The number of IMFs produced by EMD depends on the complexity of the signal and can vary from a few to many. EMD are suitable for highly non-linear and non-stationary data with highly fluctuating patterns.

There are several types of EMD methods that are used in streamflow forecasting, including classic EMD—which is the original EMD method proposed by (Huang et al., 1998), ensemble EMD (EEMD)—this method is an extension of the classic EMD that aims to reduce the effects of noise on decomposition by adding white noise to the original signal and repeating the decomposition process multiple times to generate an ensemble of IMFs, variational mode decomposition (VMD)—is based on variational calculus and aims to extract the IMFs that minimise the energy of the signal, and Complete Ensemble Empirical Mode Decomposition (CEEMD)—is a variation of EEMD which uses a modified version of shifting process to decompose the signal into IMFs, and the residue.

3.2.1 Case Studies

The study by Abudu et al. (2011) used the Transfer function noise (TFN) and ANN stochastic hybrid model to improve the accuracy of monthly streamflow forecasts during the spring-summer runoff season in Colorado. The results showed that the hybrid approach improved forecast accuracy compared to single TFN and ANN models.

In another hybrid model-based study (Kilinc & Haznedar, 2022), a combination of GA and LSTM was proposed for streamflow forecasting. The results showed that the proposed hybrid model provided better prediction accuracy results compared to the benchmark model and the linear regression model.

A hybrid model proposed by Muhammad et al. (2019a) uses LSTM and GRU for streamflow forecasting using the 35-year MOPEX data set from 10 basins in the United States. The performance of the proposed model was compared to that of conventional LSTM and GRU models and was found to outperform the LSTM model. However, the hybrid model's performance was almost the same as that of the GRU model.

4 Major Findings

- The ANNs need a cubic polynomial function to capture extreme values. The limits of the values propagated from the hidden layer to the output layer are determined by input data scaling (Imrie et al., 2000).
- The appropriate input vector for the ANN could be determined through correlation analysis (Kişi, 2007).
- Field-based soil moisture observations can be used as potential inputs in streamflow forecasting models. Studies show that including soil moisture content observations in these models improved streamflow predictions (Anctil et al., 2008). The results showed that adding a single soil moisture content time series improved prognosis by 10.7%. The study also found that the slow fluctuations of soil moisture content were more influential on streamflow predictions than the fast changes in the near-surface moisture content.
- The NEXRAD rainfall data was an excellent near-term predictor, and local meteorological variables improved 3-day forecasting but were less effective in 7- and 30-day forecasting. The SST had a long-term memory effect on streamflow prediction (Makkeasorn et al., 2008).
- The blending of large-scale circulation information and local meteorological information can benefit basin-scale streamflow forecasting (Kashid et al., 2010).
- The study demonstrates the effectiveness of LSTM in streamflow and rainfall forecasting and the improved performance of wavelet-LSTM and convolution-LSTM models compared to traditional models (Ni et al., 2020). The results indicate that WLSTM is a better alternative when the explanatory variables are small and an accurate forecast of peak values is more expected, while CLSTM is a better choice when no prior information about modelling is available and more data is available for training. It is suggested that more convolutional layers can be stacked to improve input data representation and achieve better forecast results.

- The performance of the prediction models decreases with the extension of lead time but eventually, levels off, and the inclusion of teleconnection climatic factors can improve the performance of each model, and that randomness in streamflow can weaken the prediction ability of the models. The study (Fang et al., 2023) aims to explore the evolution law of the performance of data-driven models for ten-day streamflow forecasting in the upper reaches of the Yangtze River in China. The models used in the study include MLP, SVR, FFNN and RF.
- There are definite impacts of spatial connectivity on daily streamflow prediction. The (Liu et al., 2023) proposed a new graph neural network model. This variational Bayesian edge-conditioned graph-convolution model considers the connectivity between meteorological stations to construct a graph-based basin network. This model considers the different connection relationships in the basin network using a two-dimensional edge coding method. The results showed that the proposed model outperforms other compared models regarding prediction accuracy and probabilistic prediction reliability, demonstrating the effectiveness of considering spatial connectivity.

5 Conclusions

The development and use of AI in streamflow modelling and forecasting have dramatically advanced over the past two decades. This chapter looked at the applications of AI in these areas and found that it can be highly effective when used correctly. The research done so far has shown that AI can capture the complexities of non-stationary, dynamism, non-linearity, and noise in the hydrological data. AI-based models have been shown to be effective in capturing the complex relationships between various hydrological variables, including precipitation, temperature, and runoff. These models are capable of adapting to changing conditions. They can account for the non-stationarity of hydrological processes, which are often affected by external forcings such as climate change and land use change.

Additionally, AI-based models can handle non-linear relationships and noisy data more effectively than traditional statistical models, allowing for more accurate and reliable streamflow forecasting. Overall, the use of AI-based models in hydrological research can significantly improve our understanding of water resources and facilitate better management of this critical natural resource. The reviews also found that multiple factors can impact the performance of AI modelling and the accuracy of results, including pre-processing time series frequency, selecting the most crucial input variables, and determining the most appropriate time scale. Future research has the potential to explore new and innovative approaches to streamflow forecasting. One avenue of investigation may involve integrating physical, AI, and conceptual models to achieve more accurate and reliable predictions. Additionally, there is a need to develop models that can more effectively establish the complex relationships between input variables and streamflow outputs. By exploring these new approaches, we can better understand the underlying hydrological processes and ultimately improve our ability to manage water resources sustainably and effectively.

References

Abudu, S., King, J. P., & Bawazir, A. S. (2011). Forecasting monthly streamflow of spring-summer runoff season in Rio Grande headwaters basin using stochastic hybrid modeling approach. *Journal of Hydrologic Engineering, 16*, 384–390. https://doi.org/10.1061/(asce)he.1943-5584.0000322

Akhtar, M. K., Corzo, G. A., Van Andel, S. J., & Jonoski, A. (2009). River flow forecasting with artificial neural networks using satellite observed precipitation pre-processed with flow length and travel time information : Case study of the Ganges river basin. *Hydrology and Earth System Sciences, 13*(9), 1607–1618.

Anctil, F., Lauzon, N., & Filion, M. (2008). Added gains of soil moisture content observations for streamflow predictions using neural networks. *Journal of Hydrology, 359*, 225–234. https://doi.org/10.1016/j.jhydrol.2008.07.003

Augusto, C., Santos, G., & Barbosa, G. (2014). Daily streamflow forecasting using a wavelet transform and artificial neural network hybrid models neural network hybrid models. *Hydrological Sciences Journal, 59*, 312–324. https://doi.org/10.1080/02626667.2013.800944

Besaw, L. E., Rizzo, D. M., Bierman, P. R., & Hackett, W. R. (2010). Advances in ungauged streamflow prediction using artificial neural networks. *Journal of Hydrology, 386*, 27–37. https://doi.org/10.1016/j.jhydrol.2010.02.037

Birikundavyi, S., Labib, R., Trung, H. T., & Rousselle, J. (2002). Performance of neural networks in daily streamflow forecasting. *Journal of Hydrologic Engineering, 7*, 392–398. https://doi.org/10.1061/(asce)1084-0699(2002)7:5(392)

Chu, H., Wei, J., Wu, W., et al. (2021). A classification-based deep belief networks model framework for daily streamflow forecasting. *Journal of Hydrology, 595*, 125967. https://doi.org/10.1016/j.jhydrol.2021.125967

Cigizoglu, H. K. (2005). Application of generalised regression neural networks to intermittent flow forecasting and estimation. *ASCE Journal of Hydrologic Engineering, 10*(4), 336–341.

Danandeh, A., Kahya, E., & Olyaie, E. (2013). Streamflow prediction using linear genetic programming in comparison with a neuro-wavelet technique. *Journal of Hydrology, 505*, 240–249. https://doi.org/10.1016/j.jhydrol.2013.10.003

Danandeh, A., Kahya, E., & Yerdelen, C. (2014). Computers & geosciences linear genetic programming application for successive-station monthly stream flow prediction. *Computational Geosciences, 70*, 63–72. https://doi.org/10.1016/j.cageo.2014.04.015

Dolling, O. R., & Varas, E. A. (2002). Utilisation des réseaux des neurones artificielles pour la prédiction des écoulements. *Journal of Hydraulic Research, 40*, 547–554. https://doi.org/10.1080/00221680209499899

El-Shafie, A., Abdin, A. E., Noureldin, A., & Taha, M. R. (2009). Enhancing inflow forecasting model at Aswan high dam utilising radial basis neural network and upstream monitoring stations measurements. *Water Resources Management, 23*, 2289–2315. https://doi.org/10.1007/s11269-008-9382-1

Fang, W., Zhou, J., Jia, B.-J., Gu, L., & Xu, Z.-x. (2023). Study on the evolution law of performance of mid- to long-term streamflow forecasting based on data-driven models. *Sustainable Cities and Society, 88*, 104277. https://doi.org/10.1016/j.scs.2022.104277

Firat, M., & Mahmud, G. (2007). River flow estimation using adaptive neuro fuzzy inference system. *Mathematics and Computers in Simulation, 75*, 87–96. https://doi.org/10.1016/j.matcom.2006.09.003

Huang, N. E., Shen, Z., Long, S. R., et al. (1998). The empirical mode decomposition and the Hubert spectrum for non-linear and non-stationary time series analysis. *Proceedings of the Royal Society A: Mathematical, Physical and Engineering Sciences, 454*, 903–995. https://doi.org/10.1098/rspa.1998.0193

Huang, S., Chang, J., Huang, Q., & Chen, Y. (2014). Monthly streamflow prediction using modified EMD-based support vector machine. *Journal of Hydrology, 511*, 764–775. https://doi.org/10.1016/j.jhydrol.2014.01.062

Imrie, C. E., Durucan, S., & Korre, A. (2000). River flow prediction using artificial neural networks: Generalisation beyond the calibration range. *Journal of Hydrology, 233*, 138–153. https://doi.org/10.1016/S0022-1694(00)00228-6

Ish, R. M. M., Kashid, S. S., & Ish, A. B. M. (2012). Hydrometerological modeling approaches using support vector regression (SVR) and genetic programming (GP). *ISH Journal of Hydraulic Engineering, 15*(sup1), 244–257. https://doi.org/10.1080/09715010.2009.10514978

Karenina, P., Machado, D. M., Augusto, C., et al. (2019). Analysis of the use of discrete wavelet transforms coupled with ANN for short-term streamflow forecasting. *Applied Soft Computing Journal, 80*, 494–505. https://doi.org/10.1016/j.asoc.2019.04.024

Kashid, S. S., Ghosh, S., & Maity, R. (2010). Streamflow prediction using multi-site rainfall obtained from hydroclimatic teleconnection. *Journal of Hydrology, 395*, 23–38. https://doi.org/10.1016/j.jhydrol.2010.10.004

Kilinc, H. C., & Haznedar, B. (2022). A hybrid model for streamflow forecasting in the basin of Euphrates. *Water (Switzerland), 14*. https://doi.org/10.3390/w14010080

Kişi, Ö. (2007). Streamflow forecasting using different artificial neural network algorithms. *Journal of Hydrologic Engineering, 12*, 532–539. https://doi.org/10.1061/(asce)1084-0699(2007)12:5(532)

Liu, Z., Zhou, P., Chen, G., & Guo, L. (2014). Evaluating a coupled discrete wavelet transform and support vector regression for daily and monthly streamflow forecasting. *Journal of Hydrology, 519*, 2822–2831. https://doi.org/10.1016/j.jhydrol.2014.06.050

Liu, G., Ouyang, S., Qin, H., et al. (2023). Assessing spatial connectivity effects on daily streamflow forecasting using Bayesian-based graph neural network. *Science of the Total Environment, 855*, 158968. https://doi.org/10.1016/j.scitotenv.2022.158968

Makkeasorn, A., Chang, N. B., & Zhou, X. (2008). Short-term streamflow forecasting with global climate change implications—A comparative study between genetic programming and neural network models. *Journal of Hydrology, 352*, 336–354. https://doi.org/10.1016/j.jhydrol.2008.01.023

Muhammad, A. U., Li, X., & Feng, J. (2019a). *Using LSTM GRU and hybrid models for streamflow forecasting*. Springer.

Muhammad, R., Liang, Z., Trajkovic, S., et al. (2019b). Daily stream flow prediction using optimally pruned extreme learning machine. *Journal of Hydrology, 577*, 123981. https://doi.org/10.1016/j.jhydrol.2019.123981

Mundher, Z., Ebtehaj, I., Bonakdari, H., et al. (2017). Novel approach for streamflow forecasting using a hybrid ANFIS-FFA model. *Journal of Hydrology, 554*, 263–276. https://doi.org/10.1016/j.jhydrol.2017.09.007

Ni, L., Wang, D., Singh, V. P., et al. (2020). Streamflow and rainfall forecasting by two long short-term memory-based models. *Journal of Hydrology, 583*, 124296. https://doi.org/10.1016/j.jhydrol.2019.124296

Pramanik, N., & Panda, R. K. (2009). Application of neural network and adaptive neuro-fuzzy inference systems for river flow prediction. *Hydrological Sciences Journal, 54*, 247–260. https://doi.org/10.1623/hysj.54.2.247

Ravansalar, M., Rajaee, T., & Kisi, O. (2017). Wavelet-linear genetic programming : A new approach for modeling monthly streamflow. *Journal of Hydrology, 549*, 461–475. https://doi.org/10.1016/j.jhydrol.2017.04.018

Shu, X., Ding, W., Peng, Y., et al. (2021). Monthly streamflow forecasting using convolutional neural network. *Water Resources Management, 35*, 5089–5104. https://doi.org/10.1007/s11269-021-02961-w

Sudheer, C., Maheswaran, R., Panigrahi, B. K., & Mathur, S. (2014). A hybrid SVM-PSO model for forecasting monthly streamflow. *Neural Computing and Applications, 24*, 1381–1389. https://doi.org/10.1007/s00521-013-1341-y

Toth, E., Brath, A., & Montanari, A. (2000). Comparison of short-term rainfall prediction models for real-time flood forecasting. *Journal of Hydrology, 239*, 132–147.

Valença, M., & Ludermir, T. (2001). Constructive neural networks in forecasting weekly river flow. In *Proceedings – 4th International Conference on Computational Intelligence and Multimedia Applications, ICCIMA 2001* (pp. 271–275). https://doi.org/10.1109/ICCIMA.2001.970478

Wang, Z., & Qiu, J. (2018). Hybrid models combining EMD / EEMD and ARIMA for Long-term streamflow forecasting. *Water, 2018*(7), 853. https://doi.org/10.3390/w10070853

Wang, W., Chau, K., Cheng, C., & Qiu, L. (2009). A comparison of performance of several artificial intelligence methods for forecasting monthly discharge time series. *Journal of Hydrology, 374*, 294–306. https://doi.org/10.1016/j.jhydrol.2009.06.019

Zamani, H., Phillip, J., & Abudu, S. (2017). Developing an intelligent expert system for streamflow prediction , integrated in a dynamic decision support system for managing multiple reservoirs : A case study. *Expert Systems with Applications, 83*, 145–163. https://doi.org/10.1016/j.eswa.2017.04.039

Zuo, G., Luo, J., Wang, N., et al. (2020). Decomposition ensemble model based on variational mode decomposition and long short-term memory for stream flow forecasting. *Journal of Hydrology, 585*, 124776. https://doi.org/10.1016/j.jhydrol.2020.124776

Estimation of Groundwater Levels Using Machine Learning Techniques

Sunil Gurrapu

Abstract Prediction of groundwater levels using machine learning techniques has gained substantial attention over the past few decades. Several researchers have reported the advances in this field and provide clear understanding of the state-of-the-art machine learning models implemented for GWL modelling. However, not many studies discuss the application of these results, which is of utmost importance to the practicing engineers and decision-makers for efficient planning and management of water resources infrastructure. The first part of this chapter provides a detailed review of the existing machine learning models in the domain of groundwater hydrology. This review provides a summary of all types of machine learning models developed for estimating groundwater levels over the past decade. The review also discusses the challenges associated with the use of machine learning in groundwater levels estimation. In addition, several studies from the recent past indicate the dominance of Ensemble Machine Learning in managing the sustainability of groundwater across the globe. So, the ability of ensemble machine learning models in estimating the groundwater level is discussed in the chapter. Furthermore, the chapter provides recommendations to enhance the knowledge in this domain and directions for the possible future research to improve the accuracy of the machine learning models in estimating groundwater levels.

1 Introduction

Groundwater is a vital natural resource, which plays a vital role in sustaining ecosystems, agriculture, and human societies worldwide. It is the primary source of drinking water supply to the majority of population from across the world and it is also the major contributor to irrigation sector. So, monitoring and predicting of groundwater levels is essential for sustainable water resource management, especially in regions facing water scarcity and increasing demand. To ensure sustainable

S. Gurrapu (✉)
National Institute of Hydrology, Roorkee, India

R. Srivastav, P. C. Nayak (eds.), *Applications of Machine Learning in Hydroclimatology*, The Springer Series in Applied Machine Learning,
https://doi.org/10.1007/978-3-031-64403-0_3

usage and avoid over-exploitation, efficient planning and management of groundwater resources greatly depends on the accurate prediction of groundwater levels. Accurate estimation of groundwater levels is crucial for effective water resource management, ensuring sustainable usage, and mitigating the impacts of water scarcity and overexploitation. Traditionally, these levels are estimated primarily based on the manual field observations and monitoring wells. However, estimation of groundwater levels using conventional methods alone is challenging in terms of expenditure incurred, time consumed, spatial & temporal coverage and also in terms of data quality. The emergence of machine learning techniques offers a promising solution to overcome these challenges and revolutionize groundwater level estimation (e.g. Tao et al., 2022; Bedi et al., 2020; Afzaal et al., 2020; Kenda et al., 2018).

Recent advances in soft computing techniques and/or methods have a great potential to reduce these challenges and expenditure to a great extent and enhance the accuracy of the prediction of groundwater levels. Machine learning techniques have gained popularity in groundwater modelling due to their ability to analyse complex datasets and make predictions or classifications based on patterns and relationships within the data. Several machine learning techniques developed from across the globe have emerged as valuable tools for efficient and accurate estimation of groundwater levels. A most recent publication by Tao et al. (2022) provides a clear understanding on the state-of-the-art machine learning models implemented for groundwater modelling. A range of machine learning techniques have been applied to estimate groundwater levels, with promising results. Afzaal et al. (2020) used deep learning and artificial neural networks to estimate groundwater levels, finding that the multilayer perceptron performed the best. Bedi et al. (2020) evaluated the efficacy of different machine learning models for predicting groundwater quality, finding that artificial neural networks, support vector machines, and extreme gradient boosting were all effective. In a more detailed study, Salvo (2022) highlighted that machine learning models do not offer a complete representation of the physical system, such as flux estimates or total water balance and thus cannot be used to substitute numerical models of groundwater; however, the machine learning models are affordable tools to improve predictions at a specific observation well. These studies collectively demonstrate the potential of machine learning techniques for estimating groundwater levels and quality. As machine learning continues to evolve and its applications expand, the integration of traditional methods with advanced technologies becomes increasingly important. For example, the numerical and machine learning models can be successfully used as complementary to each other, i.e. machine learning techniques can be used to improve calibration of numerical models and the results of numerical models help in understanding the physical system to select appropriate set of input variables for machine learning models (Salvo, 2022). Kenda et al. (2018) also found that with the use of strongly correlated physical parameters as input variables, machine learning models could predict groundwater level changes. With the increased access to technology and computational resources, machine learning can be harnessed as a powerful tool for sustainable groundwater level estimation and the effective management of this precious natural resource for future generations.

This chapter explores the application of machine learning techniques in estimating groundwater levels, starting with an overview of the challenges faced in traditional methods and the potential benefits of machine learning-based approaches. The first part of this chapter provides a detailed overview of the challenges faced in the traditional methods of groundwater level estimation, followed by the review of existing machine learning models in the domain of groundwater hydrology. This review provides a summary of all types of machine learning models developed for estimating groundwater levels over the past decade, i.e. during 2012 and 2022. The review also provides a list of performance criteria to choose the best machine learning model from the existing models to meet the objectives of a newly proposed study. In addition, several studies from the recent past indicate the dominance of Ensemble Machine Learning in managing the sustainability of groundwater across the globe (e.g. Manna & Anitha, 2023; Afan et al., 2021). So, the ability of ensemble machine learning models in estimating groundwater levels is also discussed in the chapter. Furthermore, the chapter provides recommendations to enhance the knowledge in this domain and directions for the possible future research to improve the accuracy of the machine learning models in estimating groundwater levels.

2 Challenges in Traditional Groundwater Level Estimation

The conventional approaches for the estimation of groundwater levels involve a great amount of manual field observations and operation and maintenance of observation wells in the region of interest. Although these approaches provide valuable data, they are associated with several challenges including the expenditure and time incurred, spatial and temporal coverage, data quality and reliability, etc. (Knowles, 2002). Traditional approach involves setting up a large number of monitoring wells in the region of interest for better estimation of the groundwater levels; however, this exercise incurs a lot of expenditure and consumes large amount of time. In effect, the monitoring wells are setup sparsely across the region and are unevenly distributed. Consequently, lack of adequate data points makes it challenging to obtain comprehensive understanding of groundwater dynamics at regional scale. This limitation of spatial coverage can greatly impact the accuracy of the groundwater level estimations, which in turn restricts the ability of the decision-maker to develop effective strategies for managing the available water resources and address the specific needs of the diverse geographical regions. In addition, the estimation of groundwater levels faces several challenges including uncertainties in input and output components, the need for long-term data for reliable recharge estimates, and the influence of specific yield uncertainty on recharge estimation (e.g. Penner et al., 2023; Addisie, 2022).

In conventional methods, the manual measurements are often taken periodically, sometimes at uneven intervals, which results in major limitation of temporal resolution. The knowledge of short-term fluctuations and rapid changes in groundwater levels is vital for modelling the groundwater dynamics and the infrequent sampling of the monitoring wells fails in capturing such fluctuations. Lack of knowledge on such important fluctuations or trends in the groundwater prevents the timely detection of extreme hydrological events including floods and droughts. In order to address any challenges associated with fluctuations in groundwater to make informed decisions and implement timely interventions, an accurate and up-to-date information is vital. However, the vulnerable areas often lack the necessary infrastructure and safety conditions to support conventional monitoring of groundwater. In addition, logistical challenges arise in the remote or inaccessible areas, leaving these regions poorly understood in terms of groundwater dynamics.

In addition to the limitations associated with temporal and spatial coverage, the quality and reliability of the data obtained through conventional methods are prone to various errors, particularly in challenging terrains or during adverse weather conditions. Human errors in addition to the equipment malfunctioning lead to inaccuracies in the recorded measurements. In addition, variations in measurement techniques and instruments used across the region can introduce inconsistencies. Such inconsistencies lead to difficulties in comparison or integration of data from various sources and undermine the confidence in the accuracy of traditional groundwater level estimations. In addition to these operational challenges, establishing a network of monitoring wells is a resources-intensive task and incurs high costs associated with drilling and installation. Moreover, regular maintenance of these wells too requires substantial amount of financial resources. This becomes a huge challenge especially in the regions of limited funding and resources, which results in inadequate installation of wells and restricts the access of decision-makers to reliable groundwater information.

Despite the limitations and challenges, traditional methods of groundwater level estimation have been the backbone of water resources management for several decades. However, the traditional methods, several times, fail to capture the complex interactions between the multitude of factors affecting the groundwater levels, including precipitation, land-use, geological formations, and human activities. So, the ability to provide a comprehensive understanding of groundwater behaviour is very much limited in the conventional methods. All these limitations in the traditional methods have led to a search of more advanced method to estimate groundwater levels and the machine learning techniques offer promising opportunities to address these limitations by integrating diverse datasets, improving prediction accuracy, and providing real-time monitoring capabilities. Developing a multi-faceted approach that combines the strengths of traditional methods and modern technologies can lead to more effective water resource management strategies that address the unique challenges faced by different regions from across the world.

3 Overview of Machine Learning Techniques

Over the past decade, machine learning techniques have gained popularity in groundwater modelling due to their ability to analyse complex datasets and make predictions or classifications based on patterns and relationships within the data (e.g. Salvo, 2022; Tao et al., 2022). Machine learning methods are closely related to data-driven models, in which a computer model learns from the observed data and makes predictions or decisions. Several machine learning methods are commonly used, each suiting different types of problem and/or data, a few commonly used ML methods include (i) Supervised Learning, (ii) Unsupervised Learning, (iii) Reinforcement Learning, (iv) Semi-supervised Learning, (v) Transfer Learning, and (vi) Deep Learning. The algorithm of supervised learning is trained on labelled data, where both input features and corresponding output labels are known. In this method, the input features are mapped with the output labels, which enables the algorithm to make predictions on unseen data (Caruana & Niculescu-Mizil, 2006). A few commonly used supervised learning algorithms include (a) Decision Trees, (b) Random Forests, (c) Support Vector Machine, (d) Naïve Bayes, and (e) Neural Networks.

Decision Tree is a non-parametric supervised learning method used for classification and regression. This method is preferred mostly for solving Classification problems. It is a tree-like model, where internal nodes represent the features of a dataset, branches represent the decision rules and each leaf node represents the outcome. Random forest belongs to the supervised learning technique and can be used for both classification and regression problems in machine learning. Random forests are ensemble of decision trees, on various subsets of the given datasets, that combine their predictions to make more accurate and robust predictions. Instead of relying on one decision tree, the random forest takes the prediction from each tree takes the prediction from each tree and based on the majority votes of prediction, it predicts the final output. The greater number of trees in the forest leads to higher accuracy and prevents the problem of overfitting. Support vector machine or SVM is one of the most popular supervised learning algorithms which can be used for both classification and regression problems. However, it is more popular for classification problems in machine learning. The goal of the SVM is to find a best decision boundary, called a hyperplane, that can segregate n-dimensional space into different classes, maximising the margin between classes. SVM chooses the extreme points/vectors, called support vectors, that help in creating the hyperplane.

Naïve Bayes classifier algorithm is a supervised learning algorithm based on Bayes theorem and is used for solving classification problems. It is a probabilistic classifier, which means it predicts on the basis of the probability of an object. Naïve Bayes assumes independence between features and calculates the probability of a certain class given the feature values. Neural Networks are composed of interconnected modes (neurons) organized in layers. They learn complex mappings by adjusting weights between nodes during training. They consist of a group of interconnected nodes, representing the neurons of a biological brain. The basic neural network consists of (i) the input layer, (ii) the hidden layer, and (iii) the output layer.

Unsupervised learning algorithm works with unlabelled data and aims to discover patterns, structures, or relationships within the data (Berry et al., 2020). Unsupervised learning methods include (a) Clustering, (b) Dimensionality Reduction, and (c) Anomaly Detection. Clustering is an unsupervised learning algorithm that groups similar data points together based on the similarity measures. There are different types of clustering algorithms based on density, distribution, centroid, and hierarchical clustering. Examples of clustering algorithm include k-means clustering and hierarchical clustering. The method of dimensionality reduction reduces the dimensionality of high-dimensional data while preserving important information. This is done for a variety of reasons, i.e. to reduce the complexity of a model, to improve the performance of a learning algorithm, or to make it easier to visualize the data. There are several techniques for dimensionality reduction, including principal component analysis (PCA), singular value decomposition (SVD), linear discriminant analysis (LDA), t-distributed stochastic neighbour embedding (t-SNE), etc. Anomaly detection methods identify unusual or anomalous patterns in data, which may indicate potential outliers or abnormal behaviour.

Reinforcement learning involves an agent that learns to interact with an environment and make decisions based on feedback or rewards. The agent takes actions in the environment and receives rewards or penalties, allowing it to learn optimal strategies (Wiering & van Otterlo, 2012). Reinforcement learning has been successful in tasks such as game playing, robotics, and control systems. The method of semi-supervised learning combines labelled and unlabelled data to perform learning tasks (Reddy et al., 2018). It leverages the additional unlabelled data to improve model performance and generalization. Transfer learning enables models trained on one task to be used as a starting point for another related task. By leveraging knowledge from the pre-trained model, transfer learning can help improve performance and reduce training time for new tasks. Dep learning is a subfield of machine learning that focuses on using deep neural networks with multiple hidden layers to learn hierarchical representations of data (Shinde & Shah, 2018). Deep learning has been particularly successful in image recognition, natural language processing, and speech recognition tasks. It is important to note that the selection of the appropriate machine learning method depends on the nature of the problem, the available data, and the desired outcome. It often requires experimentation and iterative refinement to find the best approach for a specific task.

4 Application in Groundwater Level Estimations

The application of machine learning in groundwater level estimation has gained traction across various research and practical domains. Some of the key applications include drought monitoring and prediction, flood prediction, prediction of groundwater levels, groundwater resource management, contaminant transport and remediation, aquifer recharge assessment, identification of influential factors, real-time monitoring and adaptive management, and uncertainty analysis. For example, machine learning methods are widely used for drought hazard monitoring, prediction and forecasting of drought conditions (e.g. Lalika et al., 2024; Prodhan et al., 2022). This information is crucial for drought preparedness and water allocation planning. In the context of groundwater level estimation, machine learning algorithms can process large and complex datasets, identify patterns, and build predictive models that capture the interactions between groundwater levels and various environmental factors such as rainfall, temperature, pumping rates, and soil properties. The role of machine learning in groundwater estimation is multifaceted, offering several advantages over traditional methods. By incorporating diverse environmental variables, machine learning models can capture complex interactions and identify non-linear relationships that may be overlooked by conventional methods.

Traditional groundwater monitoring methods, such as monitoring wells, often result in sparse and uneven spatial coverage. In contrast, machine learning techniques enable the integration of data from various sources, facilitating estimation at unmonitored locations and improving spatial resolution. This capability is particularly valuable in regions with limited well networks or in areas where access to monitoring wells is challenging. In addition, these models provide continuous estimates over time, overcoming the limitations of periodic manual measurements. By capturing short-term fluctuations and rapid changes in groundwater levels, machine learning-based approaches enhance the temporal resolution, allowing for more timely detection of critical events and accurate prediction of groundwater behaviour. Machine learning models can be designed to process the incoming data and provide real-time information on groundwater levels and changes. The ability to monitor groundwater levels in real time enhances the efficiency of water resource management systems, enabling prompt interventions and adaptive strategies in response to changing hydrological conditions.

Any of the above-listed machine learning algorithms can be applied to groundwater level estimation, depending on the specific characteristics of the dataset and the desired level of accuracy. Some of the commonly used machine learning techniques in groundwater level estimation include regression models, decision trees, random forests, support vector machines, and neural networks (e.g. Sahoo et al., 2017; Condon et al., 2015; Sahoo & Jha, 2015). The choice of machine learning technique depends on the nature of the data, the complexity of the problem, and the computational resources available. Their ability to handle complex datasets, adapt to changing conditions, and deliver accurate predictions can significantly contribute to sustainable water resources management (Sahoo et al., 2017). The major advantages of using machine learning techniques in groundwater level estimation are improved accuracy, reduced computational time & costs, adaptability to changing conditions, visualization and interpretability, real-time monitoring, early warning systems, the ability to incorporate various data sources, and data integration. It's worth noting that while machine learning techniques can be valuable tools in groundwater modelling, they should be used in conjunction with traditional physically based models and expert knowledge to ensure accurate and reliable results (e.g. Salvo, 2022; Kenda et al., 2018). Additionally, the interpretation of machine learning outputs should be done with caution, as these models often provide black-box solutions that may lack transparency in the decision-making process.

5 Challenges and Future Directions

While machine learning techniques offer significant promise in groundwater level estimation and water resources management, several challenges need to be addressed for their successful implementation. Some of the major challenges associated with the use of machine learning in groundwater level estimations are associated with (a) data availability and quality, (b) model interpretability, (c) quantification of uncertainty, (d) integration with traditional methods, and (e) access to technology and expertise. Data availability and data quality are two critical aspects that significantly influence the performance and success of machine learning models. In this context, data availability refers to the accessibility and quantity of data required for training and testing machine learning algorithms, while data quality pertains to the reliability, accuracy, and relevance of the data used in the model-building process. Both data availability and data quality are essential considerations in ensuring the effectiveness and generalizability of machine learning models (Ramirez et al., 2022). Machine learning models require large and diverse datasets for training and validation. Ensuring data availability and maintaining data quality remains a critical challenge, especially in regions where data collection is limited or inconsistent. The datasets with missing data can lead to biased model performance, favouring the majority class. Techniques like oversampling, under-sampling, and using class weights can help address this issue. Ramirez et al. (2022) describe that missing data can be handled through imputation techniques using remote earth observations and Palanivianayagam and Damaševičius (2023) describes other ways of handling missing data, one of the ways is to use models that can handle missing values for maintaining data integrity. In addition, high-quality data should be accurate and reliable, free from errors and inconsistencies. Noisy data can lead to misleading patterns, negatively impacting model performance. Outliers can adversely affect the model's ability to learn patterns, and handling them appropriately is critical to maintaining data quality.

While machine learning models have shown impressive performance in various tasks, their inherent complexity often makes them challenging to interpret (Carvalho et al., 2019). Model interpretability is a critical aspect, especially in domains where decisions have significant consequences. Many advanced machine learning algorithms, such as neural networks, are considered "black boxes" due to their complexity, making it challenging to interpret their predictions. They lack transparency, and the internal workings are not easily understandable. This opacity makes it challenging to interpret how and why the model arrived at a specific prediction. Understanding the importance of each feature in the model's decision-making process is crucial, although determining which features contributed most to a specific prediction in high-dimensional datasets can be difficult. For example, deep learning architectures may have millions of parameters and complexity of these parameters makes it difficult to trace how specific inputs affect the model's output. Some models sacrifice performance and predictive accuracy to increase model interpretability and provide understandable explanations. While several interpretability techniques exist, not all methods are applicable to all types of machine learning models and the choice of method depends on the model's architecture and the specific task at hand (Carvalho et al., 2019). There is no universal standard for model interpretability, and different interpretability methods may provide varying insights. This lack of standardization can make it challenging to compare and validate interpretations across models. While machine learning model interpretability remains a challenging area, ongoing research and advancements in interpretability techniques are gradually addressing these challenges.

Uncertainty quantification is one important aspect in machine learning models as groundwater systems are inherently uncertain and complex. So, developing machine learning models that can accurately quantify uncertainty in the predictions of groundwater levels is essential for reliable water resource management (Shahri et al., 2022). Uncertainty quantification in machine learning refers to the process of estimating and representing the uncertainty associated with model predictions. Uncertainty arises from various sources, including data errors, model limitations, and the inherent stochastic nature of certain processes. By quantifying uncertainty, machine learning models can provide more reliable and informative predictions, leading to increased trust and credibility in their results. Commonly used methods for uncertainty quantification in machine learning include (i) Probabilistic models, (ii) Bayesian Inference, (iii) Ensemble Methods, (iv) Dropout Technique, (v) Monte-Carlo Dropout, (vi) Quantile Regression. By estimating and representing uncertainty, machine learning models can gain increased trust and applicability in real-world scenarios, leading to more effective and responsible use of these techniques.

Combining the strengths of both machine learning models and traditional methods of groundwater level estimation can lead to improved performance, accuracy, and efficiency, but this integration requires careful calibration and validation. However, there are inherent differences and complexities that can hinder seamless integration. Traditional methods often rely on specific data formats or assumptions that may not align with the requirements of machine learning algorithms. Integrating these methods may necessitate data preprocessing, transformation, or even feature engineering to make the data suitable for machine learning models. Traditional methods are often more transparent and interpretable than complex machine learning models, which are often considered "black boxes". Validating and evaluating integrated models becomes challenging, as traditional evaluation metrics may not directly apply to machine learning models. In addition, integrating machine learning with traditional methods may demand significant computational resources, which can be a practical limitation in some applications. The integration should consider the scalability of both approaches to achieve efficient results. Ensuring data consistency and addressing any discrepancies are critical for accurate results. Integrating machine learning with traditional methods may raise concerns related to governance, compliance, and ethical considerations. Ensuring transparency, fairness, and compliance with regulations is crucial for responsible integration.

Machine learning applications require access to computational resources and skilled personnel with expertise in data science and hydrology, which may be limited in certain regions. Shortage of computational resources can hinder the ability to use complex models, leading to reduced predictive performance. Moreover, the ability to handle big data is limited, high-resolution modelling is impossible, model interpretability is limited, selection of models is limited and require longer training times. Addressing these challenges will pave the way for a more comprehensive and integrated approach to groundwater level estimation, incorporating the strengths of both traditional methods and machine learning techniques. To overcome some of these limitations, groundwater practitioners can explore strategies such as data pre-processing to reduce data size, feature selection to prioritize essential variables, using model compression techniques, and optimizing code efficiency. Additionally, cloud computing and distributed computing resources can be leveraged to handle large datasets and computationally demanding tasks, though this may incur additional costs. The integration of machine learning models into existing water management systems requires careful consideration of the governance, infrastructure, and policies surrounding groundwater resources. Further research is needed to enhance the performance and applicability of machine learning algorithms in the groundwater level estimation. Below is the list of few areas in which further research may be needed for improving the accuracy of the models based on machine learning techniques.

(a) Acquiring high-quality, comprehensive, and long-term groundwater data is crucial for developing accurate and robust machine learning models. Research should focus on improving data collection methods, filling data gaps, and exploring new data sources, such as remote sensing and satellite observations.
(b) The selection and engineering of relevant features play a vital role in model performance. Research should investigate novel approaches to feature selection, extraction, and representation that capture essential hydrological processes and improve the interpretability of models.
(c) Comparing the performance of different machine learning algorithms and ensemble methods in groundwater level estimation can help identify the most suitable models for specific contexts. Research should evaluate the trade-offs between model complexity, computational requirements, and predictive accuracy.
(d) Developing methods to quantify and propagate uncertainties through machine learning models is crucial for decision-making and risk assessment. Research should focus on incorporating uncertainty estimation into groundwater level predictions and understanding how uncertainties affect model outputs.
(e) Investigating the applicability of transfer learning and domain adaptation techniques can aid in leveraging knowledge from related hydrological domains or regions with more data to improve groundwater level estimation in data-scarce areas.
(f) Enhancing the interpretability of machine learning models is critical for gaining insights into underlying hydrological processes. Research should explore techniques to make complex models more interpretable without sacrificing predictive accuracy.

(g) Integrating machine learning models with data assimilation techniques can enhance the utilization of observational data for groundwater level estimation, especially when dealing with sparse and irregularly distributed measurements.
(h) Integrating hydrological domain knowledge with machine learning approaches can lead to more physically meaningful models. Research should explore hybrid models that combine data-driven machine learning with physics-based hydrological models.
(i) Investigating how machine learning models perform under different hydrological conditions and how they can adapt to changing circumstances is essential for reliable groundwater level estimation over time.
(j) Developing computationally efficient algorithms and model compression techniques can enable the deployment of machine learning solutions in resource-limited settings.

Collaboration between hydrologists, data scientists, and machine learning experts is critical for advancing research in these areas. Additionally, open sharing of groundwater datasets, code, and model architectures will foster the reproducibility and scalability of research findings, ultimately benefiting the broader groundwater management community. With ongoing research and collaborative efforts, machine learning will undoubtedly play a pivotal role in shaping the future of groundwater resource management and ensuring water security in an increasingly water-stressed world.

References

Addisie, M. B. (2022). Groundwater recharge estimation using water table fluctuation and empirical methods. *H2Open Journal, 5*(3), 457–468. https://doi.org/10.2166/h2oj.2022.026

Afan, H. A., Osman, A. I. A., Essam, Y., Ahmed, A. N., Huang, Y. F., Kisi, O., Sherif, M., Sefelnasr, A., Chau, K.-w., & El-Shafie, A. (2021). Modeling the fluctuations of groundwater level by employing ensemble deep learning techniques. *Engineering Applications of Computational Fluid Mechanics, 15*(1), 1420–1439. https://doi.org/10.1080/19942060.2021.1974093

Afzaal, H., Farooque, A. A., Abbas, F., Acharya, B., & Esau, T. (2020). Groundwater estimation from major physical hydrology components using artificial neural networks and deep learning. *Water, 12*(5), 18. https://doi.org/10.3390/w12010005

Bedi, A., Samal, A., Ray, C., & Snow, D. (2020). Comparative evaluation of machine learning models for groundwater quality assessment. *Environmental Monitoring and Assessment, 192*, 776. (23 pp). https://doi.org/10.1007/s10661-020-08695-3

Berry, M. W., Mohamed, A., & Yap, B. W. (2020). *Supervised and unsupervised learning for data science*. Springer. https://doi.org/10.1007/978-3-030-22475-2

Caruana, R, & Niculescu-Mizil, A. (2006). An empirical comparison of supervised learning algorithms. *Proceedings of the 23rd international conference on Machine Learning, Pittsburgh, PA*. pp. 161–168. https://doi.org/10.1145/1143844.1143865.

Carvalho, D. V., Pereira, E. M., & Cardoso, J. S. (2019). Machine learning interpretability: A survey on methods and metrics. *Electronics, 8*(8), 832. https://doi.org/10.3390/electronics8080832

Condon, L. E., Hering, A. S., & Maxwell, R. M. (2015). Quantitative assessment of groundwater controls across major US river basins using a multi-model regression algorithm. *Advances in Water Resources, 82*, 106–123. https://doi.org/10.1016/j.advwatres.2015.04.008

Kenda, K., Čerin, M., Bogataj, M., Senožetnik, M., Klemen, K., Pergar, P., Laspidou, C., & Mladenić, D. (2018). Groundwater modelling with machine learning techniques: Ljubljana Polje Aquifer. *Proceedings, 2*, 697. (8 pp). https://doi.org/10.3390/proceedings2110697

Knowles, I. (2002). Parameter estimation in groundwater modelling. *Developments in Theoretical and Applied Mechanics, XXI*, 415–421. Accessed from https://people.cas.uab.edu/~iknowles/papers/sectam.pdf

Lalika, C., Mujahid, A. H., James, M., & Lalika, M. C. S. (2024). Machine learning algorithms for the prediction of drought conditions in the Wami River sub-catchment, Tanzania. *Journal of Hydrology: Regional Studies, 53*, 101794. https://doi.org/10.1016/j.ejrh.2024.101794

Manna, T., & Anitha, A. (2023). Deep ensemble-based approach using randomised low-rank approximation for sustainable groundwater level prediction. *Applied Sciences, 13*, 3120. https://doi.org/10.3390/app13053210

Palanivianayagam, A., & Damaševičius, R. (2023). Effective handling of missing values in datasets for classification using machine learning methods. *Information, 14*(2), 92. https://doi.org/10.3390/info14020092

Penner, G. C., Martins, R. T. A. A., Rodrigues, S., & Wendland, E. (2023). Uncertainty in groundwater recharge estimation using groundwater level fluctuation and aquifer test. *Brazilian Journal of Water Resources, 28*, e11. https://doi.org/10.1590/2318-0331.282320220113

Prodhan, F. A., Zhang, J., Hason, S. S., Sharma, T. P. P., & Mohana, H. P. (2022). A review of machine learning methods for drought hazard monitoring and forecasting: Current research trends, challenged, and future research directions. *Environmental Modelling & Software, 149*, 105327. https://doi.org/10.1016/j.envsoft.2022.105327

Ramirez, S. G., Williams, G. P., & Jones, N. L. (2022). Groundwater level data imputation using machine learning and remote earth observations using inductive bias. *Remote Sensing, 14*(21). https://doi.org/10.3390/rs14215509

Reddy, Y. C. A. P., Viswanath, P., & Reddy, B. E. (2018). Semi-supervised learning: A brief review. *International journal of Engineering & Technology, 7*(1.8), 81–85.

Sahoo, S., & Jha, M. K. (2015). On the statistical forecasting of groundwater levels in unconfined aquifer systems. *Environmental Earth Science, 73*, 3119–3136. https://doi.org/10.1007/s12665-014-3608-8

Sahoo, S., Russo, T. A., Elliot, J., & Foster, J. (2017). Machine learning algorithms for modelling groundwater level changes in agricultural regions of the U.S. *Water Resources Research, 53*(5), 3878–3895. https://doi.org/10.1002/2016WR019933

Salvo, C. D. (2022). Improving results of existing groundwater numerical models using machine learning techniques: A review. *Water, 14*(15), 2307. https://doi.org/10.3390/w14152307

Shahri, A. A., Shan, C., & Larsson, S. (2022). A novel approach to uncertainty quantification in groundwater table modelling by automated predictive deep learning. *Natural Resources Research, 31*, 1351–1373. https://doi.org/10.1007/s11053-022-10051-w

Shinde, P. P., & Shah, S. (2018). A review of machine learning and deep learning applications. *Fourth international conference on Computing Communication Control and Automation.* https://doi.org/10.1109/iccubea.2018.8697857

Tao, H., Hameed, M. M., Marhoon, H. A., Zounemat-Kermani, M., Heddam, A., Kim, A., Sulaiman, A. O., Tan, M. L., Sa'adi, Z., Mehr, A. D., Allawi, M. F., Abba, S. I., Zain, J. M., Falah, M. W., Jamei, M., Bokde, N. D., Bayatvarkeshi, M., Al-Mukhtar, M., Bhagat, S. K., Tiyasha, T., Khedher, K. M., Al-Ansari, N., Shahid, A., & Yadeen, Z. M. (2022). Groundwater level prediction using machine learning models: A comprehensive review. *Neurocomputing, 489*, 271–308. https://doi.org/10.1016/j.neucom.2022.03.014

Wiering, M., & van Otterlo, M. (2012). *Reinforcement learning—State of the art*. Springer. https://doi.org/10.1007/978-3-642-27645-3

River Discharge Forecasting in Mahanadi River Basin Based on Deep Learning Techniques

Sanjay Sharma and Sangeeta Kumari

Abstract This paper presents the applicability of deep learning techniques for river discharge forecasting. Deep learning techniques have been an area of interest for forecasting in engineering because of their forecasting accuracies. Two deep learning techniques long-short term memory (LSTM) and bidirectional long-short term memory (Bi-LSTM) have been applied and compared for forecasting river discharge data in this study. In LSTM, input flows in one direction (forward) as it has only one layer. But in Bi-LSTM, input flows in two directions (forward and backward) as it has an additional layer and the output of both layers is combined together for the final output. For the evaluation of this study, daily discharge data of the Salehbhata station in the Mahanadi River basin has been used. Two statistical measures which are Nash-Sutcliffe efficiency (NSE) and root mean square error (RMSE) were used for model evaluation. Results show that both models perform well but Bi-LSTM is slightly better than LSTM in terms of both statistical measures. These results of both models are with different values of hyperparameters. The performance of these models can be different with the same values of hyperparameters and comparison of both models with the same hyperparameters may not be ideal. From this study, it is concluded that Bi-LSTM can be an alternative approach for forecasting river discharge data.

1 Introduction

Forecasting river discharge is very important for planning and managing water resources and hazard risk assessment in most parts of the world. Most agricultural activities happen around the rivers and depend on river water for irrigation and other activities. Humans living beside rivers have a high dependency on river water for their economic activities such as irrigation of agricultural land, power generation,

S. Sharma · S. Kumari (✉)
Department of Civil Engineering, NIT Jamshedpur, Jamshedpur, India
e-mail: sangeeta.ce@nitjsr.ac.in

R. Srivastav, P. C. Nayak (eds.), *Applications of Machine Learning in Hydroclimatology*, The Springer Series in Applied Machine Learning,
https://doi.org/10.1007/978-3-031-64403-0_4

and infrastructure development (Hussain & Khan, 2020). River discharge forecasting with high accuracy can benefit these economic activities. Forecasting of river discharge becomes more important for avoiding and managing the risks of floods. Floods are the most common and devastating natural disaster which cause huge economic losses and human life losses (Ding et al., 2019). Therefore, river discharge forecasting or flood forecasting is a very important task in water resources management and finding the accurate discharges in rivers will be helpful for planning and management of water resources and reducing the economic and human life losses.

The hydrological forecasting models are usually two types: physical model and data-driven model. Physical models are mathematical models converted from hydrological cycle mechanism. These models have been used widely due to their physical basis and relatively simple model structures (Cui et al., 2021). However, physical models cannot simulate the complex situations of watersheds accurately because watersheds contain various non-linear influencing factors. Therefore, data-driven technology is used widely for forecasting models nowadays. Data-driven models mainly focus on the linear and non-linear relationship between input and output data rather than the physical mechanism of the hydrological process (Yan et al., 2021). Data-driven modelling includes machine learning models such as artificial neural network (ANN), support vector regressor (SVR), random forest (RF) and some deep learning techniques such as recurrent neural network (RNN), long short-term memory (LSTM), etc. These models have made a tremendous growth in research in forecasting modelling in recent years. These models characterize the dynamic and complex behaviour of time series data by efficient and flexible algorithms. ANN has been used as a data-driven-based hydrological forecasting model in various studies (Nourani et al., 2009; Rezaeian-Zadeh et al., 2013; Tiwari & Chatterjee, 2010). Lin et al. (2006), Nayak & Ghosh (2013) and Su et al. (2014) have used SVR for forecasting the floods and discharges in rivers. Muñoz et al. (2018), Schoppa et al. (2020) and Wang et al. (2015) have used RF for flood forecasting. Deep learning models have appeared as best reliable models for forecasting and used for flood prediction (Apaydin et al., 2020; Shakir Ali Ali et al., 2022).

The LSTM based on recurrent neural network (RNN) is a deep learning technique which has an ingenious memory cell and is designed to learn dependencies and remember information for a longer period of time. It is able to trail the dynamic and complex behaviour of the nonlinear system and can extract the features of data from different time spans (Dong & Yang, 2020). It has been used successfully for predicting the time series of hydrological parameters. LSTM model has been used in various studies for forecasting floods, streamflow, weather prediction and for simulating the groundwater level. Bidirectional LSTM (Bi-LSTM) is also the same as LSTM except input flows in both directions in the Bi-LSTM model. It is capable of capturing the information from both sides. Yan et al. (2021) used the LSTM model for streamflow prediction. Ding et al. (2020) used the LSTM model for forecasting floods. Le et al. (2019) applied LSTM for flood forecasting using daily discharge and rainfall data and found the predictive abilities of model quite impressive. Roy et al. (2022) used LSTM and Bi-LSTM for multi-step forecasting of evapotranspiration and found both models equally good as training station dataset.

Cui et al. (2021) used LSTM combined with Xinanjiang (XAJ) conceptual model for multistep ahead flood forecasting. LSTM with other models as hybrid model has been applied for flood forecasting and given acceptable outcomes (Hayder et al., 2023; Kao et al., 2020). Fu et al. (2019) used Bi-LSTM for weather prediction and found it performing well. Shakir Ali Ali et al. (2022) used the convolutional neural network (CNN)-based LSTM model for simulating the groundwater level.

Most of the machine learning models have been used for flood forecasting or streamflow forecasting. Deep learning models have great results in forecasting floods in many studies, However, Bi-LSTM has not been explored so much in river flow forecasting. This study considers the forecasting of river discharge by LSTM and Bi-LSTM models. Also, this study compares the performance of both models and explains model's parameters input value differences in both models.

2 Study Area and Dataset

Mahanadi River is one of India's largest rivers and lies in the Chhattisgarh and Odisha states of India. The origin of the river is in the Dhamtari district of Chhattisgarh state and it drains into the Bay of Bengal. Total length of the river is 851 Km and it covers the basin area of 141,589 km^2. The mean maximum and minimum temperatures in the basin are 40 °C and 20 °C. This river basin is mostly rainfed and annual average rainfall in this basin is 1572 mm. Monsoon occurs mostly from June to October in this basin. Because of uneven rainfall in this basin, some districts suffer drought also. The Mahanadi River basin is divided into three sub-basins which are upper, middle and lower basins. The coastal parts of the Mahanadi River face waterlogging problems. Figure 1 shows the study area map of the Mahanadi River basin.

This study was done on Salehbhata station which is located on Ong River. Ong River lies in the middle basin of the Mahanadi River basin and covers a total length of 215 Km to reach to Mahanadi River. Daily river discharge data for 20 years from 1999 to 2018 was taken from the India Water Resources Information System (WRIS). The mean and standard deviation of daily discharge data are 62.62 m^3/s and 283.77 m^3/s. Minimum and maximum discharge values at Salehbhata station are 0 m^3/s and 7916 m^3/s. The daily discharge data is shown in Fig. 2.

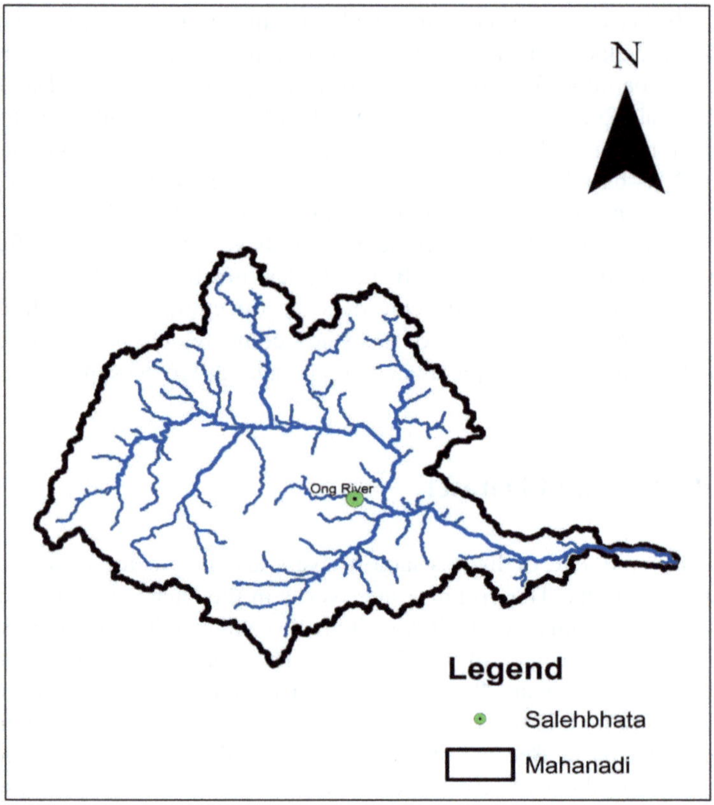

Fig. 1 Study area map of Mahanadi River Basin

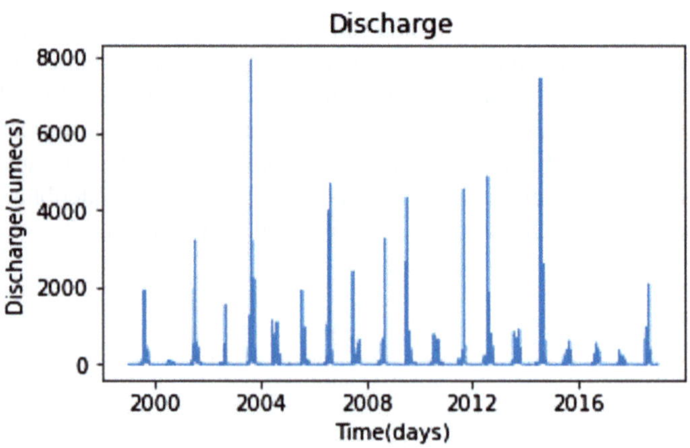

Fig. 2 Daily discharge data at Salehbhata station

3 Methodology

3.1 *Long Short-Term Memory (LSTM) and Bidirectional LSTM (Bi-LSTM)*

LSTM network is the modified version for recurrent neural network with memory recall function and was introduced by Hochreiter and Schmidhuber in 1997. The LSTM network's hidden layer not only takes information from the input layer at the same time step but also takes information from the output of the hidden layer of the previous time step. The LSTM blocks replace the hidden units of the RNN network to make it an LSTM network. These blocks contain three gates which are input gate, forget gate, and output gate. These gates control the flow of information in their memory. The LSTM network contains the key component of repeating blockchain, called as cell state, for learning from the time series data. These gates are used to predict and control the cell state. Input gate controls the information coming from the input gate of the same time step and decides which information should be kept in the cell state. Forget gate decides which information should be dropped from the cell state. Output gate decides which information should go into the output network from the cell state.

$$H_t = f(UX_t + WH_{t-1}) \tag{1}$$

$$y_t = VH_t \tag{2}$$

where U is the weight between the input layer and the hidden layer and V is the weight between the hidden layer and the output layer. W is the weight between hidden layers of current time step and previous time step. X_t and y_t are the input and output at time t. H_t is the state of the hidden layer at time t. For performing the regression task, the LSTM model uses four layers which are input layer, LSTM layer, fully connected layer and regression layer. The input layer and fully connected layer communicate with input and output variables respectively. LSTM layer helps the hidden layer and regression layer perform the regression task. LSTM layer is the important component of LSTM network as it improves the long-term dependencies within various time steps. In the LSTM network, hidden layer contains single LSTM layer and it flows in one direction (forward). While in Bi-LSTM, hidden layer contains two LSTM layers. One layer flows in the forward direction and the other flows in the backward direction. The final output of Bi-LSTM depends on these two layers. Thus Bi-LSTM captures the past and future information of data. Figure 3 shows the architecture of LSTM and Bi-LSTM models.

The performance of these models was evaluated by statistical metrics. Two statistical metrics, root mean square error (RMSE) and Nash–Sutcliffe efficiency (NSE), were used in this study. These statistical metrices have been used in many research studies in hydrology.

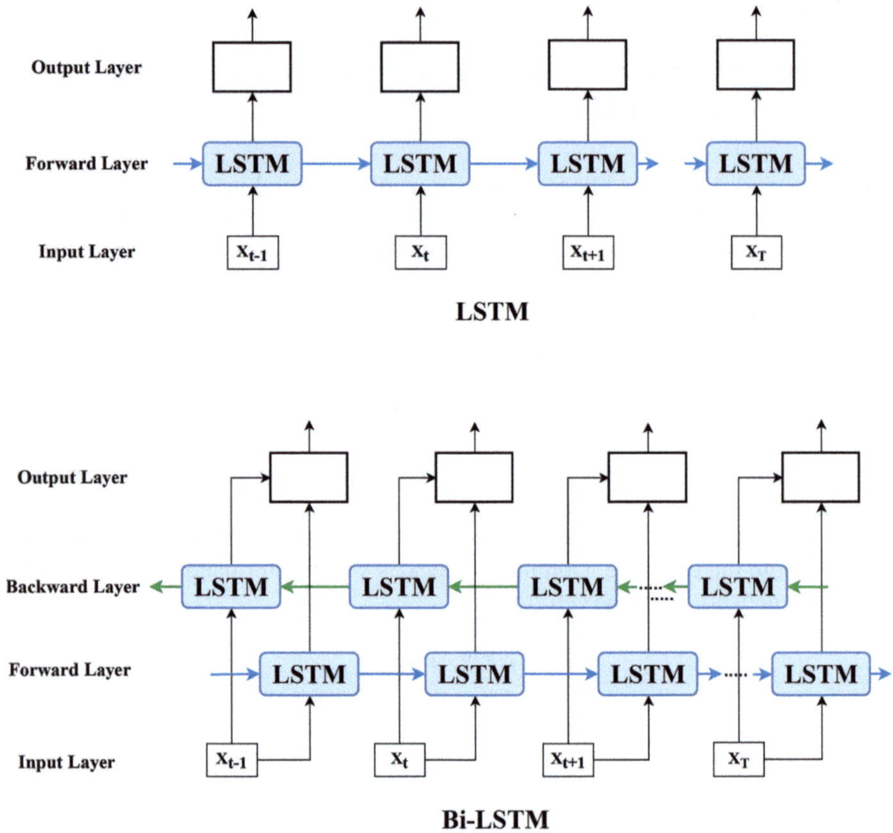

Fig. 3 Simple structure of LSTM model and Bi-LSTM model

$$\text{RMSE} = \sqrt{\frac{1}{n}\sum_{i=1}^{n}\left(Q_{\mathrm{i}} - (Q_{\mathrm{o}})_{\mathrm{i}}\right)^2} \tag{3}$$

$$\text{NSE} = 1 - \frac{\sum_{i=1}^{n}\left(Q_{\mathrm{i}} - (Q_{\mathrm{o}})_{\mathrm{i}}\right)^2}{\sum_{i=1}^{n}\left(Q_{\mathrm{i}} - q\right)^2} \tag{4}$$

where Q_{i} is the predicted value, Q_{o} is the observed value, q is the mean of observed values and n is the number of observations.

4 Results and Discussions

Both models (LSTM and Bi-LSTM) were trained and tested with daily river discharge data. Data was divided into 80% of training and 20% of testing. Time steps of 7 days was taken for prediction. Models were tested with different time steps also but with higher number of time steps, accuracy of models was reducing. Both models are optimized manually and tested with different values of parameters of models. Results show that the best performance of both models did not come with the same parameter values as shown in Table 1. Models were tested with different values of these parameters. With changing the values of batch size, both models' performances were not affected. So, the higher value of batch size was chosen as it reduces the running time of models.

The models were validated with 30% of training data and training and validation loss were calculated. Figure 4 shows that the validation losses are more than training losses but the difference is not much. The losses for both training and validation do not change after the certain value of epochs as shown in Fig. 4. This value of epochs is higher for the Bi-LSTM model than the value in the LSTM model. Validation losses stop declining before training losses in the LSTM model while the declination stops at the same time for validation and training losses in the Bi-LSTM model. So, the overfitting is more controlled in the Bi-LSTM model. Training processes had to be stopped after validation losses stops changing to control the overfitting. These losses show the reason of choosing different values of epochs for both models.

Table 1 Parameters of LSTM and Bi-LSTM models

Parameters	LSTM	Bi-LSTM
No. of layers	1	3
No. of nodes	20	10
Epochs	50	70
Batch size	128	128

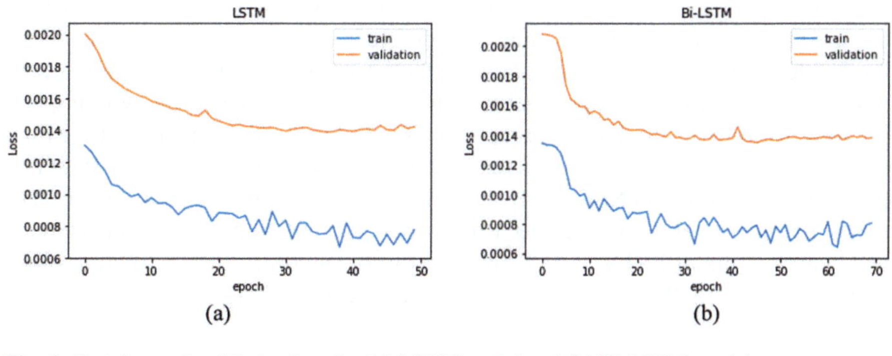

Fig. 4 Training and validation loss for (**a**) LSTM model and (**b**) Bi-LSTM model

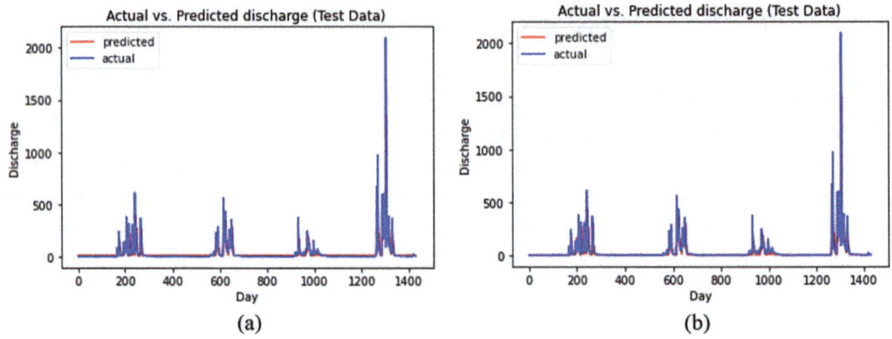

Fig. 5 Training and validation loss for (**a**) LSTM model and (**b**) Bi-LSTM model

Results showed that both models are working fine for forecasting the daily river discharge data. The RMSE values for LSTM and Bi-LSTM models are 73.62 m^3/s and 72.786 m^3/s respectively. The NSE scores for LSTM and Bi-LSTM models are 0.4641 and 0.4763 respectively. Both the evaluation models show the Bi-LSTM model is slightly better than the LSTM model. Models are not predicting high values with good accuracies. Both models need improvement for high values. Figure 5 shows the predicted and actual values for testing period for both models. There are errors in predictions but both models are able to predict the values. Bi-LSTM model is better for high altitude values which is creating the difference of accuracies of both models.

5 Conclusions

In this study, daily river discharge data was forecasted based on LSTM and Bi-LSTM models. Data was divided into 80% of training and 20% of testing. RMSE and NSE were used as evaluation models for analysing the performance of models. Models were tested with various values of model's parameters which means manually optimized. It was found from the results that best performance of both models (LSTM and Bi-LSTM) came with different values of parameters. So, it can be concluded that for better comparison of models, both models should be tested with various values of each parameter of the models. Both models are working fine for forecasting river discharge. However, both models need improvement for high values prediction which also can increase the overall model's performance. Based on the results of RMSE and NSE, the Bi-LSTM model is working better than the LSTM model for daily river discharge forecasting. So, Bi-LSTM models can be an alternate and better option for LSTM models for river discharge forecasting. For future work, both models should be compared for longer period forecasting.

Acknowledgement The authors are thankful to India Water Resources Information System (WRIS) for providing the data freely through the online site.

References

Apaydin, H., Feizi, H., Sattari, M. T., Colak, M. S., Shamshirband, S., & Chau, K. W. (2020). Comparative analysis of recurrent neural network architectures for reservoir inflow forecasting. *Water (Switzerland), 12*(5). https://doi.org/10.3390/w12051500

Cui, Z., Zhou, Y., Guo, S., Wang, J., Ba, H., & He, S. (2021). A novel hybrid XAJ-LSTM model for multi-step-ahead flood forecasting. *Hydrology Research, 52*(6), 1436–1454. https://doi.org/10.2166/NH.2021.016

Ding, Y., Zhu, Y., Wu, Y., Jun, F., & Cheng, Z. (2019). Spatio-temporal attention LSTM model for flood forecasting. In *2019 international conference on Internet of Things (IThings) and IEEE Green Computing and Communications (GreenCom) and IEEE Cyber, Physical and Social Computing (CPSCom) and IEEE Smart Data (SmartData)* (pp. 458–465). https://doi.org/10.1109/iThings/GreenCom/CPSCom/SmartData.2019.00095

Ding, Y., Zhu, Y., Feng, J., Zhang, P., & Cheng, Z. (2020). Interpretable spatio-temporal attention LSTM model for flood forecasting. *Neurocomputing, 403*, 348–359. https://doi.org/10.1016/j.neucom.2020.04.110

Dong, W., & Yang, Q. (2020). Data-driven solution for optimal pumping units scheduling of smart water conservancy. *IEEE Internet of Things Journal, 7*(3), 1919–1926. https://doi.org/10.1109/JIOT.2019.2963250

Fu, Q., Niu, D., Zang, Z., Huang, J., & Diao, L. (2019). Multi-stations weather prediction based on hybrid model using 1D CNN and Bi-LSTM. In *2019 Chinese Control Conference (CCC).*

Hayder, I. M., Al-Amiedy, T. A., Ghaban, W., Saeed, F., Nasser, M., Al-Ali, G. A., & Younis, H. A. (2023). An intelligent early flood forecasting and prediction leveraging machine and deep learning algorithms with advanced alert system. *PRO, 11*(2). https://doi.org/10.3390/pr11020481

Hussain, D., & Khan, A. A. (2020). Machine learning techniques for monthly river flow forecasting of Hunza River, Pakistan. *Earth Science Informatics, 13*(3), 939–949. https://doi.org/10.1007/s12145-020-00450-z

Kao, I. F., Zhou, Y., Chang, L. C., & Chang, F. J. (2020). Exploring a long short-term memory based encoder-decoder framework for multi-step-ahead flood forecasting. *Journal of Hydrology, 583*, 124631. https://doi.org/10.1016/j.jhydrol.2020.124631

Le, X. H., Ho, H. V., Lee, G., & Jung, S. (2019). Application of long short-term memory (LSTM) neural network for flood forecasting. *Water (Switzerland), 11*(7). https://doi.org/10.3390/w11071387

Lin, J. Y., Cheng, C. T., & Chau, K. W. (2006). Using support vector machines for long-term discharge prediction. *Hydrological Sciences Journal, 51*(4), 599–612. https://doi.org/10.1623/hysj.51.4.599

Muñoz, P., Orellana-Alvear, J., Willems, P., & Célleri, R. (2018). Flash-flood forecasting in an andean mountain catchment-development of a step-wise methodology based on the random forest algorithm. *Water (Switzerland), 10*(11). https://doi.org/10.3390/w10111519

Nayak, M. A., & Ghosh, S. (2013). Prediction of extreme rainfall event using weather pattern recognition and support vector machine classifier. *Theoretical and Applied Climatology, 114*(3–4), 583–603. https://doi.org/10.1007/s00704-013-0867-3

Nourani, V., Komasi, M., & Mano, A. (2009). A multivariate ANN-wavelet approach for rainfall-runoff modeling. *Water Resources Management, 23*(14), 2877–2894. https://doi.org/10.1007/s11269-009-9414-5

Rezaeian-Zadeh, M., Tabari, H., & Abghari, H. (2013). Prediction of monthly discharge volume by different artificial neural network algorithms in semi-arid regions. *Arabian Journal of Geosciences, 6*(7), 2529–2537. https://doi.org/10.1007/s12517-011-0517-y

Roy, D. K., Sarkar, T. K., Kamar, S. S. A., Goswami, T., Muktadir, M. A., Al-Ghobari, H. M., Alataway, A., Dewidar, A. Z., El-Shafei, A. A., & Mattar, M. A. (2022). Daily prediction and multi-step forward forecasting of reference evapotranspiration using LSTM and Bi-LSTM models. *Agronomy, 12*(3). https://doi.org/10.3390/agronomy12030594

Schoppa, L., Disse, M., & Bachmair, S. (2020). Evaluating the performance of random forest for large-scale flood discharge simulation. *Journal of Hydrology, 590*(September), 125531. https://doi.org/10.1016/j.jhydrol.2020.125531

Shakir Ali Ali, A., Ebrahimi, S., Masood Ashiq, M., Alasta, M. S., & Azari, B. (2022). CNN-Bi LSTM neural network for simulating groundwater level. *Computational Research Progress in Applied Science & Engineering, 8*(1), 1–7. https://doi.org/10.52547/crpase.8.1.2748

Su, J., Wang, X., Liang, Y., & Chen, B. (2014). GA-based support vector machine model for the prediction of monthly reservoir storage. *Journal of Hydrologic Engineering, 19*(7), 1430–1437. https://doi.org/10.1061/(asce)he.1943-5584.0000915

Tiwari, M. K., & Chatterjee, C. (2010). Development of an accurate and reliable hourly flood forecasting model using wavelet-bootstrap-ANN (WBANN) hybrid approach. *Journal of Hydrology, 394*(3–4), 458–470. https://doi.org/10.1016/j.jhydrol.2010.10.001

Wang, Z., Lai, C., Chen, X., Yang, B., Zhao, S., & Bai, X. (2015). Flood hazard risk assessment model based on random forest. *Journal of Hydrology, 527*, 1130–1141. https://doi.org/10.1016/j.jhydrol.2015.06.008

Yan, L., Chen, C., Hang, T., & Hu, Y. (2021). A stream prediction model based on attention-LSTM. *Earth Science Informatics, 14*(2), 723–733. https://doi.org/10.1007/s12145-021-00571-z

Machine Learning Models for Groundwater Level Prediction

Mayank Raturi, Deepak Khare, and Nitesh Patidar

Abstract Groundwater is considered as cleanest and widely used source of water throughout the world. Proper management of groundwater resources requires continuous monitoring of groundwater level. To make decisions and plans for future, decision-makers need reliable tools to predict the groundwater level by developing relationships between the factors/parameters that can affect the dynamics of groundwater. With the effects of climate change such as increasing heat, higher rainfall and frequently occurring extreme weather events such as floods and draughts, the need of tools and models for studying the effects of changing climate on groundwater level fluctuations has increased. Such tools and models help decision-makers in better decision-makings. With the recent development in the field of artificial intelligence, the reliable predictions of groundwater levels are possible using machine learning models. Machine learning models are data-driven models which can produce useful results utilizing the diverse data sets and computational infrastructure. This chapter focuses on types of machine learning models such as Artificial Neural Network (ANN), Adaptive Neuro Fuzzy Inference System (ANFIS), Support Vector Mechanism (SVM). Random Forest (RF) and Decision Trees. A detailed review of the algorithm of each model and their application in groundwater level prediction is discussed. A brief literature review of groundwater level prediction using machine learning approaches is presented for better understanding of the topic.

M. Raturi · D. Khare
Department of Water Resources Development and Management, Indian Institute of Technology, Roorkee, India

N. Patidar (✉)
National Institute of Hydrology, Roorkee, India
e-mail: npatidar.nihr@gov.in

R. Srivastav, P. C. Nayak (eds.), *Applications of Machine Learning in Hydroclimatology*, The Springer Series in Applied Machine Learning,
https://doi.org/10.1007/978-3-031-64403-0_5

1 Introduction

The development of techniques and models that allow computer systems to learn from data and gradually improve their performance on a given task without being explicitly programmed constitutes the field of machine learning, a branch of artificial intelligence. In other words, machine learning algorithms can automatically recognize patterns, trends, and relationships in massive amounts of data that may be utilized to make predictions or choices. Natural language processing, image recognition, fraud detection, and recommendation systems are just a few examples of the many applications that can be made use of with machine learning algorithms. These algorithms can be trained on a variety of data types, including images, text, and numerical data.

Because machine learning can offer a few benefits over conventional mathematical models, it has grown in popularity in the water resources and hydrology. Here are a few justifications for why machine learning is being applied to water resources.

Handling Complex and Non-linear Relationships Machine learning algorithms are well-suited to handling complex and non-linear relationships between input and output variables, which is often the case in water resources management. Traditional mathematical models may be limited in their ability to capture these relationships, which can lead to inaccurate predictions and decisions.

Dealing with Large and Diverse Data Machine learning algorithms can handle large and diverse datasets, which is becoming increasingly important as more data is collected through remote sensing and other sources. Traditional mathematical models may struggle to handle these large datasets and may require simplifications or assumptions that can limit their accuracy.

Scalability and Adaptability Machine learning algorithms can be trained on large datasets and can be easily adapted to new data and changing conditions. This makes them more scalable and adaptable than traditional mathematical models, which may require significant time and resources to modify or update.

Improved Accuracy Machine learning algorithms can often provide improved accuracy and prediction performance compared to traditional mathematical models. This is especially true in cases where the underlying relationships between variables are complex and non-linear.

Overall, machine learning can provide several advantages over traditional mathematical models in water resources management, including the ability to handle complex relationships, large datasets, and changing conditions, as well as improved accuracy and prediction performance.

2 Machine Learning Models

In the field of water resource management, decision-makers require a tool that can deliver in less time with less inputs. Traditional models require many input parameters and become very complex for a large data set which leads to difficulties in model parameter estimation. Also, to obtain reliable results from such models distributed and accurate inputs are required which are often not available. To solve this problem researchers in different parts of the world pushed themselves to explore machine learning. Following are some of the widely used models for simulation and prediction in the field of groundwater:

2.1 Artificial Neural Networks (ANNs)

An artificial neural network (ANN) model is a data-driven model which works like a biological neural network in the human brain. The first neural network model was developed by McCulloch and Pitts in 1943. An ANN consists of a variable number of elements known as neurons. An ANN model typically consists of three layers: input, hidden, and output layers (Fig. 1). Each layer consists of neurons with similar properties. Input layers take input variables (e.g., rainfall, temperature, groundwater level). A relative weight is given to each input parameters depending upon the effect of parameters on the output. In the hidden layers, each neuron adds all the inputs and then applies an activation function to calculate the output. The output layers consist of outputs produced by the hidden layer.

ANN calculates or simulates the results by learning the way the system behaves. For that, ANN relates each input parameter to each output parameter which is called the training process. Training of the model means to change the architecture of ANN to enhance network efficiency. It includes tuning the adjustable parameters like weights assigned to input, creating or pruning the connections (Kwon, 2011).

2.1.1 Architecture

In this block diagram, you can see the basic architecture of a feedforward neural network, which is the most common type of artificial neural network. Data is fed into the network at the input layer. The data is then passed through one or more hidden layers, each of which contains multiple neurons. Each neuron in the hidden layer gets information from neurons in the layer below and sends its output to neurons in the layer above. The output layer then creates the network's ultimate output (Banerjee et al., 2011).

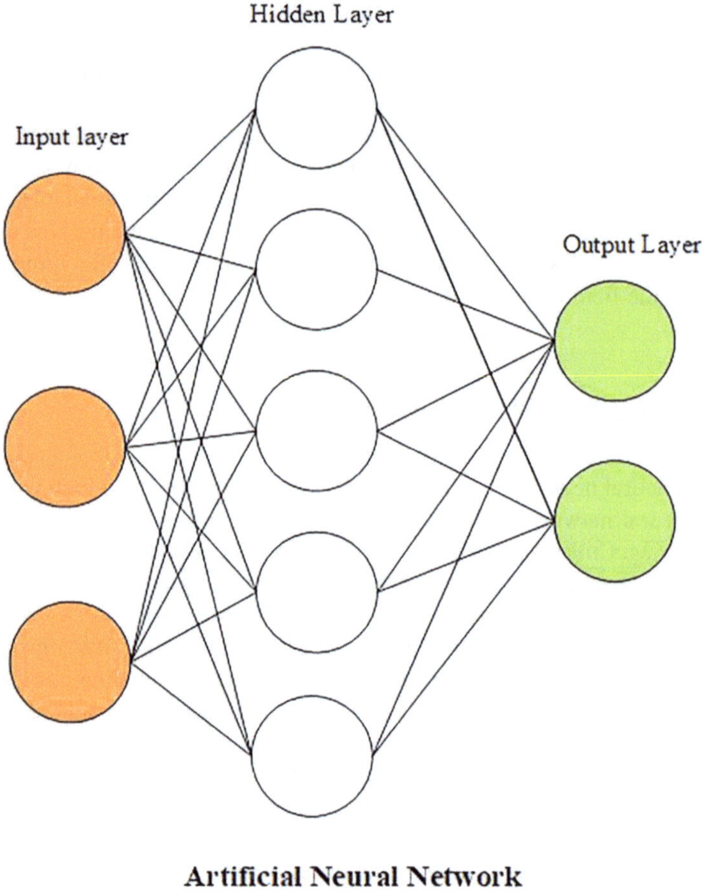

Fig. 1 Architecture of artificial neural network

2.1.2 Working

Artificial neural networks (ANNs) are a type of machine learning algorithm that are roughly based on the physiology and function of the human brain. They are used to recognize patterns in data, make predictions, and perform various other tasks. Here's a general overview of how an ANN works (Banerjee et al., 2011):

1. *Inputs are fed into the network*: These inputs could be anything from images to numerical data.
2. *Weights are applied to the inputs*: Each input is multiplied by a weight (w_i), which determines how important that input is to the output of the network.
3. *Activation function is applied*: The weighted inputs are passed through an activation function (f(net)), which helps to determine the output of the network. The

variable net in the activation function is the scaler product of input and weight applied to it.

$$\text{net} = w^T x = w_1 x_1 + \ldots + w_n x_n$$

4. *Outputs are generated*: The outputs are the result of the calculations performed by the activation function.

$$O = f(\text{net}) = f\left(\sum_{i=1}^{n} w_i x_i\right)$$

5. *The network is trained*: The network modifies the weights and biases of its neurons during training in order to minimize the discrepancy between the output it predicts and the output that occurs.

2.2 Adaptive Neuro-Fuzzy Inference System (ANFIS)

ANFIS stands for Adaptive Neuro-Fuzzy Inference System. It is a type of artificial neural network that integrates the principles of fuzzy logic and neural networks to create a hybrid system that can perform both classification and regression tasks. ANFIS was first introduced by Jang in 1993. ANFIS has several applications in various fields, including engineering, finance, medicine, and robotics. Here are some common uses of ANFIS (Miro et al., 2021):

Prediction: ANFIS can be used to predict outcomes based on input variables. For example, it can be used to predict stock prices, weather forecasts, or energy consumption.

Control: ANFIS can be used to control systems based on input variables. For example, it can be used to control the temperature of a room based on the outside temperature and the number of people in the room.

Classification: ANFIS can be used to classify data based on input variables. For example, it can be used to classify emails as spam or not spam based on their content.

Optimization: ANFIS can be used to optimize processes based on input variables. For example, it can be used to optimize the performance of a manufacturing process by adjusting the input parameters.

Diagnosis: ANFIS can be used to diagnose medical conditions based on input variables. For example, it can be used to diagnose heart disease based on a patient's age, blood pressure, and cholesterol levels.

Robotics: ANFIS can be used to control robots based on input variables. For example, it can be used to control the movement of a robot arm based on the position of an object.

In summary, ANFIS can be used for a wide range of applications that involve prediction, control, classification, optimization, diagnosis, and robotics. Its versatility and ability to learn from data make it a powerful tool for solving complex problems in various fields.

2.2.1 Architecture

The architecture of ANFIS is based on five layers (Fig. 2), each with a specific function (Miro et al., 2021):

Layer 1 (Input Layer): this layer is a fuzzy layer in which each node i is an adaptive node. B and C are the input of the system. O_i is the output of the ith node if layer 1. Each adaptive node is a square function which is represented by the following equation:

$$O_{1,i} = \mu_{B,i}(B) \text{ for } i = 1, 2$$

$$O_{1,j} = \mu_{C,i}(C) \text{ for } j = 1, 2$$

where $\mu_{B,\ i}$ and $\mu_{C,\ i}$ are membership functions.

Layer 2 (Fuzzification Layer): this layer checks weights allotted to each membership function. It receives input from layer 1 as B_i and C_i. every node in this layer is fixed and output is the product of all the incoming signals. The output of this layer can be represented as

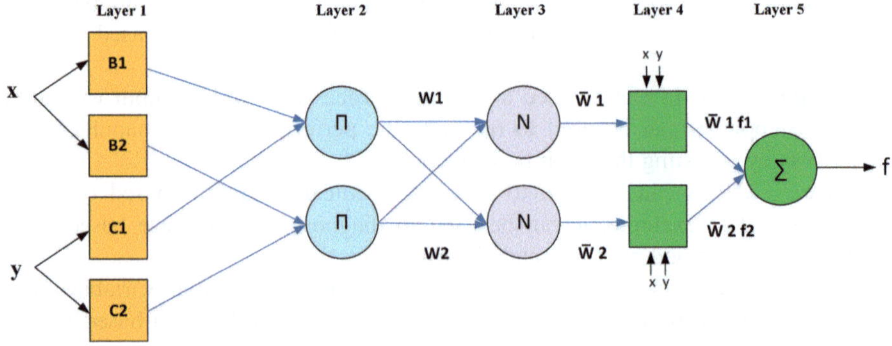

Fig. 2 Architecture of ANFIS

$$O_{2,i} = w_i = \mu_{B,i} \cdot \mu_{C,j} \; i = 1, 2$$

Layer 3 (Rule Layer): each node in this layer is marked with a circle labelled with *N*, which indicates normalization to the output of the previous layer. The *i*th node in this layer calculates the ratio of the weight of *i*th node to the sum of all the weight. The output of this layer can be expressed as:

$$O_{3,i} = \frac{w_i}{w_1 + w_2} \quad i = 1, 2$$

Layer 4 (Inference Layer): This layer provides output *y* resulting from the inference of rule. The output of this layer is simply the product of normalized weights and first-order polynomial.

$$O_{4,i} = \overline{w_i} f_i = \overline{w_i}(p_i B + q_i C + r_i) \; i = 1, 2$$

Layer 5 (Output Layer): This layer is called as output layer which sums all the input coming from layer 4. This layer consists of a single node labelled as $\sum$. The output of this layer is expressed as:

$$O_{5,i} = \sum_{i=0}^{n} \bar{w_i} f_i = \frac{\sum_{i=0}^{n} \bar{w_i} f_i}{w_1 + w_2}$$

The ANFIS architecture uses a hybrid learning algorithm that combines the backpropagation algorithm and the least squares method. The backpropagation algorithm is used to adjust the parameters in layers 2 and 3, while the least squares method is used to adjust the parameters in layer 5. This hybrid learning algorithm allows ANFIS to learn from data and improve its accuracy over time.

In summary, ANFIS is a hybrid system that combines the principles of fuzzy logic and neural networks to create a powerful tool for classification and regression tasks. Its architecture consists of five layers, each with a specific function, and uses a hybrid learning algorithm to improve its accuracy.

2.2.2 Working

The working of ANFIS can be understood in the following steps:

1. *Data pre-processing*: The first step is to pre-process the data by cleaning and transforming it into a format suitable for ANFIS. This may involve normalizing the data, removing outliers, and converting categorical variables into numerical form.

2. *Fuzzy inference system*: ANFIS uses a fuzzy inference system (FIS) to model the relationship between the input variables and the output variable. The FIS consists of a set of rules that map input variables to output variables based on their linguistic values (e.g. "high", "low", "medium"). The FIS is trained using a hybrid learning algorithm that combines backpropagation and gradient descent methods.
3. *Neural network*: ANFIS also includes a neural network component that learns to adjust the parameters of the FIS based on the input data. The neural network is trained using a supervised learning algorithm, such as gradient descent or Levenberg-Marquardt algorithm.
4. *Testing and validation*: Once the ANFIS model is trained, it can be used to forecast new data by inputting the values of the input variables and calculating the output variable based on the FIS and neural network parameters. The accuracy of the model can be validated using various techniques such as cross-validation, hold-out validation, or time-series forecasting.

2.3 Support Vector Mechanism (SVM)

A well-liked machine learning approach for classification and regression analysis is called Support Vector Machine (SVM). Vapnik's SVM proposal was based on statistical theory. For both linear and non-linear data analysis, it is a potent tool. SVM operates by locating the ideal hyperplane that divides the data into various classes. A line or plane known as a hyperplane divides the data into two groups according to their characteristics. The best hyperplane is that which maximizes the margin, where margin is the separation between the hyperplane and the nearest data points for each class. The main principle of SVM is to use a kernel function to convert the input data into a high-dimensional feature space. The input data are mapped by the kernel function into a higher-dimensional space where a hyperplane can be used to divide the data more easily. The input variables are combined linearly to represent the hyperplane in this space.

SVM finds the optimal hyperplane by solving a quadratic optimization problem that maximizes the margin subject to certain constraints. The constraints ensure that the data is correctly classified and that the margin is maximized.

In addition to its use in classification, SVM can also be used for regression analysis. In this case, by reducing the sum of the squared errors, SVM determines the hyperplane that matches the data the most effectively. Compared to other machine learning methods, SVM provides a few benefits. In high-dimensional spaces, it works well, can handle non-linear data, and is less prone to overfitting. It is widely utilized in a variety of industries, including bioinformatics, computer vision, and natural language processing (Noble, 2006).

2.3.1 Architecture

The architecture of Support Vector Machine (SVM) is a supervised learning algorithm that is used for classification and regression analysis (Fig. 3). The main components of the SVM architecture are as follows (Noble, 2006):

Input data: The input data consists of a set of observations or examples, where each observation is represented by a set of features or attributes.

Kernel function: The kernel function is a mathematical function that transforms the input data into a higher-dimensional feature space. The kernel function can be linear, polynomial, radial basis function (RBF), sigmoid, or custom. An SVM estimator f on regression can be expressed as:

$$f(x) = w.\varnothing(x) + b$$

where w is the weight vector, b is bias, and $\varnothing(x)$ represents a nonlinear transfer function.

Decision boundary: The decision boundary is a hyperplane that separates the input data into different classes. The decision boundary is chosen to maximize the margin, which is the distance between the hyperplane and the closest data points from each class.

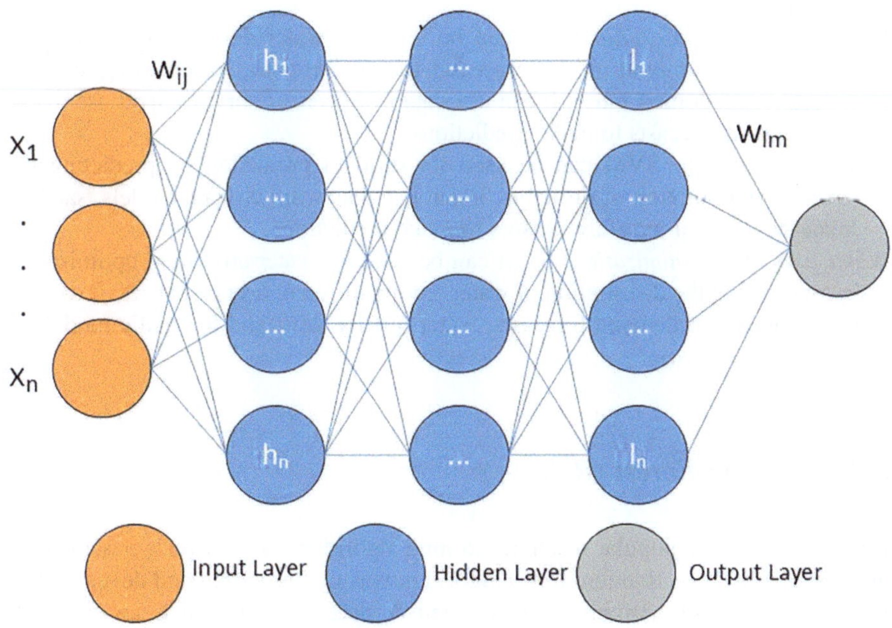

Fig. 3 Architecture of SVM

Support vectors: The support vectors are the data points that are closest to the decision boundary. These data points define the margin and play a critical role in the optimization problem that SVM solves.

Optimization problem: Vapnik introduced an optimization problem with an ε insensitivity loss function to optimize the SVM estimator.

$$\min \frac{1}{2}\|w\|^2 + C\sum_{i=1}^{m}(\xi + \xi_i)$$

Output: The output of SVM is a binary or multi-class classification label for each observation or a continuous value for regression analysis.

Support Vector Machine (SVM) has several applications in water resources and hydrology. Here are some common applications of SVM in this field:

Flood forecasting: SVM can be used for flood forecasting by predicting the water level or discharge of a river. It can use data from weather forecasts, satellite imagery, and stream gauges to make predictions.

Groundwater modelling: SVM can be used for groundwater modelling to predict the water table elevation or the flow direction of groundwater. It can use data from well logs, geophysical surveys, and water quality tests to make predictions.

Water quality prediction: SVM can be used for water quality prediction by predicting the concentration of pollutants in rivers, lakes, and reservoirs. It can use data from water quality tests, weather forecasts, and land use maps to make predictions.

Water demand forecasting: SVM can be used for water demand forecasting by predicting the amount of water needed for different uses, such as agriculture, industry, and domestic use. It can use data from census surveys, land use maps, and weather forecasts to make predictions.

Drought prediction: SVM can be used for drought prediction by predicting the severity and duration of droughts. It can use data from climate models, satellite imagery, and soil moisture sensors to make predictions.

Water allocation optimization: SVM can be used for water allocation optimization by optimizing the distribution of water among different users and uses. It can use data from water demand forecasts, water quality tests, and hydraulic models to make predictions.

2.4 Random Forest (RF)

Random Forest is a popular machine learning algorithm used for both classification and regression tasks. Random forest algorithm was first proposed and developed by L. Breiman in 2001 (Breiman, 2001; Biau & Scornet, 2016). It is an ensemble method that combines multiple decision trees to make predictions or decisions.

The basic idea behind a random forest is to create a set of decision trees, each trained on a random subset of the data, and then combine their predictions to produce a more accurate and reliable result. Individual decision trees in a random forest are learned from different subsets of the training data that are randomly sampled and replaced, i.e., bootstrapping. In addition, only a random subset of features (variables) is considered for sharing at each decision tree node, which helps reduce the risk of overfitting and improves the generalization of the model. Predictions from individual trees in a random forest are combined to produce a final prediction in a process called "voting" for classification tasks or averaging for regression tasks. Random Forests are known for their ability to handle high-dimensional data, handle missing values, and provide good accuracy in various applications such as image recognition, text classification, and financial forecasting. They are also robust to outliers and noisy data, making them a popular choice for many real-world machine learning problems.

2.4.1 Algorithm

There can be many meaning to the term random forest. For some researchers, it is a general expression for computing. A random forest predictor consists of M randomized regression trees. For the jth tree in the tree family, the predicted value at point x is denoted by m_n $(x;\ \Theta_i,\ D_n)$, where $\Theta_1,\ldots,\ \Theta_M$ are independent random variables. The jth tree estimate can be present mathematically.

$$m_n(X; \Theta_j, D_n) = \sum_{i \in D_n(\Theta_j)} \frac{X_i \in A_n(x : \theta_j, D_n) Y_i}{N_n(X : \theta_j, D_n)}$$

where $D_n(\Theta_j)$ is the set of data points selected prior to the tree construction, $A_n(X;\ \Theta_j,\ D_n)$. $A_n(X;\ \Theta_j,\ D_n)$ is the cell containing x, and $N_n(X;\ \Theta_j,\ D_n)$ is the number of (preselected) points that fall into $A_n(X;\ \Theta_j,\ D_n)$.

Data Preparation Collect and pre-process the groundwater level data, which typically entails gathering historical measurements of the groundwater level and related features (such as precipitation, temperature, well location, etc.) to produce a feature matrix X (representing the input features) and a target vector Y (representing the groundwater level measurements).

Random Subset Selection (Bootstrapping) Create several subsets of the training data (x, y) by selecting subsets at random and utilizing replacement with bootstrapping.

Decision Tree Creation Create a decision tree for each subgroup using a random subset of X's features at each node to partition the data. This is accomplished by repeatedly separating the data into the relevant characteristics and values up until a stopping requirement (such as reaching a maximum depth, a minimum number of samples per leaf, etc.) is satisfied.

Ensemble Creation To generate a collection (ensemble) of decision trees, repeat the previous step numerous times (often several hundred or thousand times). In order to add randomness and variety to the ensemble, each decision tree is constructed using a separate subset of the training data and a different selection of characteristics.

Prediction By averaging the anticipated values from each tree, the ensemble of decision trees can be used to estimate groundwater levels in regression. The final anticipated groundwater level is taken to be the average prediction.

Model Evaluation On a validation or test dataset, assess the effectiveness of the Random Forest model using the proper evaluation metrics, such as mean squared error (MSE), root mean square error (RMSE), mean absolute error (MAE), etc. This aids in evaluating the model's precision and generalizability.

Hyperparameter Tuning You can choose to use methods like cross-validation or grid search to improve the Random Forest model's hyperparameters, such as the number of trees in the forest, the maximum depth of the trees, the number of features to take into account at each node, etc. The performance of the model can be further enhanced by tuning hyperparameters.

Prediction on New Data After being trained, improved, and assessed, the Random Forest model can be used to forecast new, unforeseen data, such as estimating groundwater levels at various locations and times.

3 Applications of Machine Learning Models in Groundwater

With advancement in the computational techniques, many machine learning models have been used for forecasting groundwater parameters. Some of the major applications of ML models in groundwater hydrology are explained below:

3.1 Groundwater Level Prediction

Groundwater level is an important source of water for domestic, agriculture, and industrial purposes. For proper management of groundwater, accurate assessment of groundwater level is very important. In the last few decades, there has been an increase in the use of ML models for groundwater level prediction as compared to numerical model like MODFLOW. The main reason behind it was that ML models are easy to use and require a smaller number of input parameters. Chen et al. (2020) use ML models SVM to simulate groundwater level in the Heihe River basin, China. RMSE and r^2 were used to evaluate the accuracy of the model. Some of the results are shown below in the figure. This study finds that SVM performed best in training and validation stages. It concluded that with correct choice of algorithm, SVM can be a good option for simulating GWL.

Jeihouni et al. (2019) used ANN to simulate groundwater level in aquifers with limited data. The study region was the alluvial plain of Chamchamal in Iran, which is encircled by a karstic structure. In order to train the ANN model, hundreds of data sets were created using the calibrated MODFLOW simulation of groundwater flow. Another goal was to find ANN models that, even with only a small amount of training data, could simulate the complicated dynamics of GWL. Different ANN models were used, with various combinations of input data, to accomplish this goal. Additionally, various network topologies were assessed, each with a different number of hidden layers and activation algorithms. This study concluded that the ANN model can be effectively used for GWL simulation.

In the research of Coppola et al. (2003) ANNs are employed to precisely anticipate transient water levels in a complicated groundwater system under varying aquifer stressors. The Upper Floridian aquifer, which is made up of high permeability karst limestone overlain by a low permeability semiconfining unit and is topped by a surficial sandy unconfined aquifer, served as the model's test site in the Northwest Hillsborough Wellfield near Tampa Bay, Florida, USA. Using root mean square error and absolute mean error, the outcomes of numerical and machine learning models were compared for typical monitoring wells. Compared to the numerical flow model, ANN was far more accurate at simulating the oscillation of the water levels. During each stress period, the absolute mean error of the numerical model was greater than the greatest ANN prediction error for any given observation. The authors came to the conclusion that for some situations, ANN represents a better alternative to numerical modelling methodologies since it doesn't call for conditions that are time- and space-variable and difficult to describe in aquifers.

3.2 Groundwater Quality Prediction

With the availability of large data sets related to quality of groundwater, machine learning models are widely used for water quality prediction. This can help decision-makers to solve the problem in less time. These models can consider factors such as land use, soil type, and pollution sources to predict the levels of different contaminants in the groundwater.

Palani et al. (2008) used ANN model to predict water quality parameters (salinity, temperature, dissolved oxygen) using weekly observed data of water quality at different locations. The study shows that the GRNN and Ward Net architecture shows the best performance based on different water quality variables.

Latif et al. (2020) used ANN model with back propagation algorithm which represents a non-linear relationship to estimate the total nutrient concentration in a reservoir in Taiwan. R^2 were used to evaluate the performance of the model. The study concluded that ANN is a good model to predict water quality parameters.

As a method for managing coastal aquifers, Banerjee et al. (2011) compared the application of ANN simulation to mathematical modelling. The models' goals were to predict the increase in groundwater salinity brought on by pumping at various rates on the island of Kvaratti in the Lakshadweep archipelago (India) and to identify management solutions to prevent the increase in groundwater salinity. SUTRA was used to create a 2D finite element model that is physically based. By comparing root mean squared error (RMSE) and mean absolute error (MAE), the study showed that ANN outperformed the physically-based model.

3.3 Groundwater Recharge Estimation

Machine learning models can be used to estimate the amount of water that enters the groundwater system through recharge. These models can consider factors such as precipitation, land use, and soil type to estimate the recharge rate.

Martinsen et al. (2022) used machine learning model based on random forest regressor to map groundwater recharge coefficient using a range covariate related to geology, soil, topography and climate. The groundwater recharge map provides high-resolution estimates of Europe and nearby areas where information is otherwise not available.

4 Conclusion

This chapter examined several machine learning models for predicting groundwater levels, highlighting their architecture, advantages, and applications. Since groundwater is a crucial worldwide resource, it must be managed and monitored effectively. However, traditional approaches like numerical modelling frequently have trouble handling big, complicated datasets with non-linear correlations. In tackling these issues, machine learning methods such as Random Forest (RF), Support Vector Machine (SVM), Adaptive Neuro-Fuzzy Inference System (ANFIS), and Artificial Neural Networks (ANN) provide notable benefits. These models offer enhanced accuracy and prediction performance because they can handle intricate interactions, sizable and varied datasets, and adapt to changing circumstances. Using these sophisticated models, decision-makers can estimate groundwater levels with more accuracy, improving groundwater resource management and sustainability overall.

References

Banerjee, P., Singh, V. S., Chatttopadhyay, K., Chandra, P. C., & Singh, B. (2011). Artificial neural network model as a potential alternative for groundwater salinity forecasting. *Journal of Hydrology, 398*(3–4), 212–220. https://doi.org/10.1016/j.jhydrol.2010.12.016

Breiman, L. (2001). Random forests. *Machine Learning, 45*, 5–32. Kluwer Academic Publishers. https://doi.org/10.1023/A:1010933404324

Biau, G., & Scornet, E. (2016). A random forest guided tour. *TEST, 25*(2), 197–227. https://doi.org/10.1007/s11749-016-0481-7

Chen, C., He, W., Zhou, H., Xue, Y., & Zhu, M. (2020). A comparative study among machine learning and numerical models for simulating groundwater dynamics in the Heihe River Basin, northwestern China. *Scientific Reports, 10*(1), 1–13. https://doi.org/10.1038/s41598-020-60698-9

Coppola, E., Szidarovszky, F., Poulton, M., & Charles, E. (2003). Artificial neural network approach for predicting transient water levels in a multilayered groundwater system under variable state, pumping, and climate conditions. *Journal of Hydrologic Engineering, 8*(6), 348–360. https://doi.org/10.1061/(asce)1084-0699(2003)8:6(348)

Jeihouni, E., Eslamian, S., Mohammadi, M., & Zareian, M. J. (2019). Simulation of groundwater level fluctuations in response to main climate parameters using a wavelet–ANN hybrid technique for the Shabestar Plain, Iran. *Environmental Earth Sciences, 78*(10). https://doi.org/10.1007/s12665-019-8283-3

Kwon, S. J. (2011). Artificial neural networks. *Artificial Neural Networks*, 1–426. https://doi.org/10.15864/jmscm.1104

Latif, S. D., Azmi, M. S. B. N., Ahmed, A. N., Fai, C. M., & El-Shafie, A. (2020). Application of artificial neural network for forecasting nitrate concentration as a water quality parameter: A case study of Feitsui Reservoir, Taiwan. *International Journal of Design and Nature and Ecodynamics, 15*(5), 647–652. https://doi.org/10.18280/ijdne.150505

Martinsen, G., Bessiere, H., Caballero, Y., Koch, J., Collados-Lara, A. J., Mansour, M., Sallasmaa, O., Pulido-Velazquez, D., Williams, N. H., Zaadnoordijk, W. J., & Stisen, S. (2022). Developing a pan-European high-resolution groundwater recharge map – Combining satellite data and national survey data using machine learning. *Science of the Total Environment, 822*, 153464. https://doi.org/10.1016/j.scitotenv.2022.153464

Miro, M. E., Groves, D., Tincher, B., Syme, J., Tanverakul, S., & Catt, D. (2021). Adaptive water management in the face of uncertainty: Integrating machine learning, groundwater modeling and robust decision making. *Climate Risk Management, 34*, 100383. https://doi.org/10.1016/j.crm.2021.100383

Noble, W. S. (2006). What is a support vector machine? *Nature Biotechnology, 24*(12), 1565–1567. https://doi.org/10.1038/nbt1206-1565

Palani, S., Liong, S. Y., & Tkalich, P. (2008). An ANN application for water quality forecasting. *Marine Pollution Bulletin, 56*(9), 1586–1597. https://doi.org/10.1016/j.marpolbul.2008.05.021

Genetic Algorithm-Aided Neural Network for Sediment Critical Shear Stress Modeling

Umesh K. Singh, Pijush Dutta, and Sanjeet Kumar

Abstract Critical shear stress (CSS) of sediment governs its transport in open channel flow. Earlier, mathematical models were developed to compute CSS of sediment under clay influence which was limited to their own experimental data. The present study aims to develop an ANN model to compute CSS of coarser sediment present in mobile channel bed made of cohesive sediment mixture. The proposed model was optimized using parent-selected operator GA (genetic algorithm), which helped to determine the optimal process parameters responsible for CSS. Initially, the mathematical model was implemented with the help of ANN (artificial neural network) and later on it was optimized by three well-known parent-selected operator genetic algorithms. Data from the literature along with current experimental data was used to develop ANN-based model to compute CSS for coarser particle under clay influence. It was found that linear ranked selected GA-ANN is the best-fitted model for both training and testing data. The CSS obtained for optimized input values for clay fraction by weight (CP), weighted geometric standard deviation of sediment mixture (SM), and dimensionless dry bulk unit weight of cohesive sediment (DB) are 0.104, 2.9523, and 1.6921 respectively which were quite satisfactory and performing better than the existed mathematical models.

U. K. Singh (✉)
Water Resources Systems Division, National Institute of Hydrology Roorkee, Roorkee, India
e-mail: ukrsingh.nihr@gov.in

P. Dutta
Department of Electronics & Communication Engineering, Greater Kolkata College of Engineering & Management, Baruipur, West Bengal, India

S. Kumar
Department of Civil Engineering, Koneru Lakshmaiah Education Foundation Deemed to be University, Guntur, Andhra Pradesh, India

R. Srivastav, P. C. Nayak (eds.), *Applications of Machine Learning in Hydroclimatology*, The Springer Series in Applied Machine Learning,
https://doi.org/10.1007/978-3-031-64403-0_6

1 Introduction

The transport of sediment in open channel leads to aggradation and degradation issues that cause several problems like failure of hydraulic structures across and along the channel, flood control, and reservoir sedimentation. For example, failure of Islam weir on Sutlej River and lowering of Yellow River in China due to degradation (Garde & RangaRaju, 2000). Aggradation leads to deposition of sediments in reservoir that causes reduction in water storage capacity of reservoirs. For example, the storage capacity of Nizam Sagar (India) was lost about 60.7% over a period of 62 years, whereas the corresponding loss in Maithon reservoir in Jharkhand was about 55% over 50 years reported by Rathore et al. (2006). The transport of sediment from mobile bed in open channel occurs when the shear stress developed by incoming flow on the sediments present on channel bed just exceeds the critical shear stress of sediment on that channel bed. Hence, critical shear stress plays an instrumental role in transport of sediment. The present study focused on the critical shear stress in the presence of clay in the sediment mixture of channel bed. The model to compute CSS in cohesive sediment is significantly different from cohesionless sediment due to cohesive properties of clay (Singh & Ahmad, 2019). The research has been done well to compute CSS for the cohesionless sediment (Shields, 1936, Singh et al., 2019); however; it is under investigation for cohesive mixture of the channel bed. Kothyari and Jain (2008) developed a model to compute CSS of coarser particle for clay-sand mixture and clay-sand-gravel mixture as Eq. (1) along with supporting Eq. (2).

$$(\tau_{cc}/\tau_{cm}) = 0.94(1 + P_c)^{3/2 e_r^{-1/6}(1+0.001\mathrm{UCS}^*)^{9/20}} \tag{1}$$

$$\mathrm{UCS}^* = \mathrm{UCS}/(\rho_S - \rho) g d_a \tag{2}$$

Here, τ_{cc} is the critical shear stress of cohesive sediment mixture (N/m^2); τ_{cm} the Shields critical shear stress for cohesionless sediment having size equal to arithmetic mean size of cohesive sediment mixture (N/m^2); P_c the clay percentage in fraction by weight; e_r the void ratio of cohesive sediment mixture; UCS^* the dimensionless unconfined compressive strength of cohesive sediment mixture; and UCS is the unconfined compressive strength of cohesive sediment mixture (N/m^2); ρ_s and ρ the particle and fluid densities (kg/m^3) respectively; g the gravitational acceleration (m/s^2); d_a is arithmetic mean size of the cohesive sediment mixture (m). Singh et al. (2017) proposed the model to compute CSS of gravel particles for clay-silt-gravel mixture and clay-silt-sand-gravel mixture as Eq. (3) along with supporting Eq. (4).

$$(\tau_{cc}/\tau_{cm}) = 1 + 0.1428 P_c^{1.294} \sigma_{cc}^{0.553} (\gamma_b/\gamma_w)^{3.814} \tag{3}$$

$$\sigma_{cc} = \frac{\sum\left(d_{50}{}^{*}\sigma_{g}\right)}{d_{a}} \quad (4)$$

Here, σ_{cc} is the weighted geometric standard deviation of the cohesive sediment mixture, γ_b the bulk unit weight (N/m^3) of the cohesive sediment mixture, γ_w the unit weight of water (N/m^3), and d_{50} the median size of sediment particle.

Ahmad et al. (2018) proposed a model to compute CSS of gravel particles for clay-silt-gravel mixture and clay-silt-sand-gravel mixture as Eq. (5).

$$(\tau_{cc}/\tau_{cm}) = 1 + 0.354P_c^{2.081}(1 + P_s)^{2.275}(\gamma_b/\gamma_w)^{4.10} \quad (5)$$

Here, P_s is the silt percentage in fraction by weight.

Singh et al. (2022) predicted CSS for gravel particles in cohesive mixture by using machine learning; however, they didn't incorporate clay-silt-sand mixture in their study. The present study aims to fill the gap to compute CSS based on ANN by considering the present experimental data on clay-silt-sand mixture along with the previous data from Kothyari and Jain (2008), Singh et al. (2017), and Ahmad et al. (2018). The study incorporated the existing CSS model from Kothyari and Jain (2008), Singh et al. (2017), and Ahmad et al. (2018) (i.e., Eq. (1), (3), and (5)) to compare them with proposed ANN model.

2 Proposed Approach for Modelling

Exploration of critical shear stress or incipient motion study is significantly important in sediment transport in open channel. The present study, along with experimental investigation, considered a new artificial neural network (ANN) optimized genetic algorithm (GA-ANN) model to determine critical shear stress (CSS) of coarser particles in a cohesive mixture with variables clay percentage (in fraction) by weight (CP), weighted geometric standard deviation of sediment mixture (SM), and dimensionless dry bulk unit weight of cohesive sediment (DB). All these three input variables are considered as input nodes of the artificial neural network model. To obtain the coefficient of the model an evolutionary algorithm called genetic algorithm (GA) was used. In the present study, 148 experimental datasets were used where 118 datasets were used for the cross-validation and training, and the rest of the 30 datasets were used for testing datasets. Initially, the best mathematical was identified which fitted with the experimental datasets with higher accuracy. After this, the optimal point of critical shear stress (CSS) of coarser particles in a cohesive mixture was found out with variables clay percentage (in fraction) by weight (CP), weighted geometric standard deviation of sediment mixture (SM), and dimensionless dry bulk unit weight of cohesive sediment (DB). To obtain the optimal condition, the resulting objective functions of the mathematical model are obtained by linear equation implemented by CP, SM & DB and finally, it is optimized by GA.

2.1 Neural Network Model for Critical Shear Stress (CSS)

It is considered that the critical shear stress (CSS) is the linear function of a mixture of clay percentage (in fraction) by weight (CP), weighted geometric standard deviation of sediment mixture (SM), and dimensionless dry bulk unit weight of cohesive sediment (DB). A simple neuron without any hidden layer is used to represent a linear function. So, it can be written as Eq. (6)

$$\mathrm{CSS} = f_1(\mathrm{CP}, \mathrm{SM}, \mathrm{DB}) \tag{6}$$

For each of these neurons, nodes of the input layer are CP, SM, and DB whereas critical shear stress (CSS) is the node of the output layer. Their weights, i.e., w_1,w_2, w_3 are associated with the three inputs respectively, and bias term β is associated with the output node. Here, no activation function is not a classification problem. So, according to the neural network model critical shear stress (CSS) can be expressed as follows:

$$\mathrm{CSS} = \mathrm{CP} * w_1 + \mathrm{SM} * w_2 + \mathrm{DB} * w_3 + \beta \tag{7}$$

Now, the optimal values of weights and bias values of Eq. (7) are determined by the optimization techniques. For this purpose, genetic algorithm as an optimization algorithm is used to find out the neural network structure of functions CSS. The proposed ANN model was designed with the help of training datasets obtained from the experiments. From all the datasets of 148, 80% of the data (118) was used for training the proposed model while 20% of the data (30) was for testing or new cases for validation purposes.

2.2 Parent Selection Operator of GA (PSO-GA)

Parent selection is the process of selecting parents which mate and recombine to create offspring for the next generation. Parent selection is very crucial to the convergence rate of the GA to give a better and fittest solution. However, avoiding one extremely fit solution over the whole population in a few generations leads to a loss of diversity. To maintain good diversity, multiple fit solutions for the entire population are considered. Three different parent selection operators (PSO) of GA are considered, namely: the roulette wheel selection (RWS), the linear rank selection (LRS), and the tournament selection (LS). Table 1 shows the value of parameter setting for GA-ANN modeling.

Table 1 Parameter setting for GA for ANN modeling

Parameters of GA	Value
No. of population (n_{pop})	100
No. of parameters to be optimized	4
Range of CP	[0.1, 0.5]
Range of SM	[2.917781, 7.490329]
Range of DB	[1.49376, 2.025484]

2.2.1 The Roulette Wheel Selection (RWS)

The salient characteristic of this operator is the fact that it gives to each individual i of the current population a probability $p(i)$ of being selected, proportional to its fitness $f(i)$ shown in Eq. (8):

$$p(i) = \frac{f(i)}{\sum_{j=1}^{n} f(i)} \tag{8}$$

where n denotes the population size in terms of the number of individuals. The RWS can be implemented according to the following pseudo-code:

```
Calculate the sum S = Σ_{j=1}^{n} f(i)
while   for each individual 1 ≤ i ≤ n //do Generate a random number α ∈ [ 0, S]
                      iSum = 0; j = 0
        while         iSum (<α and j < n ) do iSum ← iSum + f(i)
                      j ← j + 1
        end
                      Select the individual j
end
```

The main drawback of this technique is premature convergence of the GA to the local optimum.

2.2.2 Linear Rank Selection (LRS)

LRS is also a variant of RWS that aims to fix the problem of the premature convergence of the GA to a local optimum. Instead of considering an individual's fitness, it bases decisions on their rank. The best person receives a number of n, while the worst person receives a rank of 1. Each person i has the likelihood of getting chosen given by the expression as Eq. (9) dependent on its rank.

$$(i) = \frac{\text{rank}\ (i)}{n * (n-1)} \tag{9}$$

Once all individuals of the current population are ranked, the linear ranked selection (LRS) algorithm can be implemented according to the following pseudo code:

```
Calculate the sum v = 1/(n − 2.00)
do    For each individual 1 ≤ i ≤ n
      Generate a random number α ∈ [0, v]
      do    For each 1 ≤ j ≤ n
            do    If (p(j) ≤ α )
                  Select the jth individual
                  Break
            end
      end
end
```

2.2.3 Tournament Selection (TS)

Tournament selection is a variant of rank-based selection operators. Its principle consists of randomly selecting a set of k individuals. These individuals are then ranked according to their relative fitness and the fittest individual is selected for reproduction. The whole process is repeated n times for the entire population. Hence, the probability of each individual to be selected is given by Eq. (10):

$$p(i) = \begin{cases} \dfrac{C_{n-1}^{k-1}}{C_n^k} & \text{if } i \in [1, n-k-1] \\ 0; & \text{otherwise} \end{cases} \tag{10}$$

The implementation of TS can be performed according to the below pseudo-code.

```
Create a table t where the n individuals are placed in a randomly chosen order
do    for i = 1 to n
                                  Generate a random number i1 ∈ [1, n]
      do          for j = 1 to n
                                  Generate a random number i2 ∈ [1, n]
                                  With i2 ≠ i1
                                  If f(i1) > f(i2) select i1 else select i2
      end         end for j
end   end for i
```

3 Result Analysis

The numerical simulation results for the modeling of critical shear stress (CSS) of coarser particles in cohesive mixture were done using three popular Parent Selection Operators of GA (PSO-GA). In addition, comparisons among the performance of the proposed algorithms and statistical analysis of the evaluated results have been carried out. For all cases, initial population and maximum iteration number are set to 1000 and 5000 respectively. For ANN-based model, search space is restricted to four, i.e., six-dimensional function is considered for the optimization problem to search optimal values of the coefficients $\{w_1, w_2, w_3, \beta\}$. The search range has been set to [− 15, 15] for all coefficients of both types of modeling.

All the techniques are simulated using MATLAB 2013b in a computer with 16 GB RAM, 11th Generation, core TM i5 processor, and Windows 10 Operating System. Due to the stochastic nature of meta-heuristics, different outputs may be obtained depending on different random initializations. Therefore, each algorithm is executed 10 times for each case and the statistical analysis has been carried out from the obtained results. During these numerical simulations, the efficiency of the proposed algorithms has been tested and compared with the efficiency of the proposed algorithms on the basis of some criteria such as fitness test, computational efficiency, convergence test, and accuracy test. Table 2 represents the coefficient of all the proposed PSO-GA-ANN models obtained after executing the algorithms.

Figures 1, 2, and 3 represent the cost value of three proposed algorithms GA-RWS-ANN, GA-LRS-ANN, and GA-TS-ANN respectively. From the graphical plot the least cost value obtained by the Roulette wheel parent-selected GA-optimized ANN model has been observed.

Relative error is one of the most important statistical parameters which are mostly used for the identification of the best-fitted model. It can be defined as the ratio of individual absolute error to the experimental output. Relative error for all the three algorithms is shown in Fig. 4. Eq. (11) and (12) represent the mathematical formulation of relative error (RE).

$$\text{RE} = \frac{IAE}{F_{\text{Measured}}} \tag{11}$$

$$\text{RE} = \frac{|F_{\text{measured}} - F_{\text{calculated}}|}{F_{\text{measured}}} \tag{12}$$

where F_{measured} = measured value & $F_{\text{calculated}}$ = calculated value.

Table 2 Coefficient of PSO-GA-ANN algorithms

Name of the algorithm	w_1	w_2	w_3	β
RWS-GA-ANN	3.4745	0.1881	−0.4667	1.0610
LRS-GA-ANN	4.1155	0.2161	3.515	−6.0187
TS-GA-ANN	3.8260	0.2035	1.7589	−3.0555

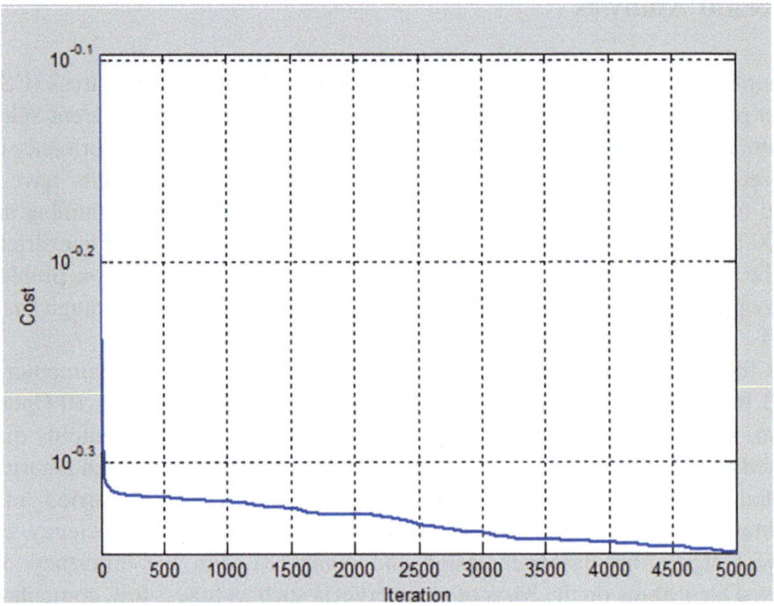

Fig. 1 Objective function of GA-RWS-ANN algorithm

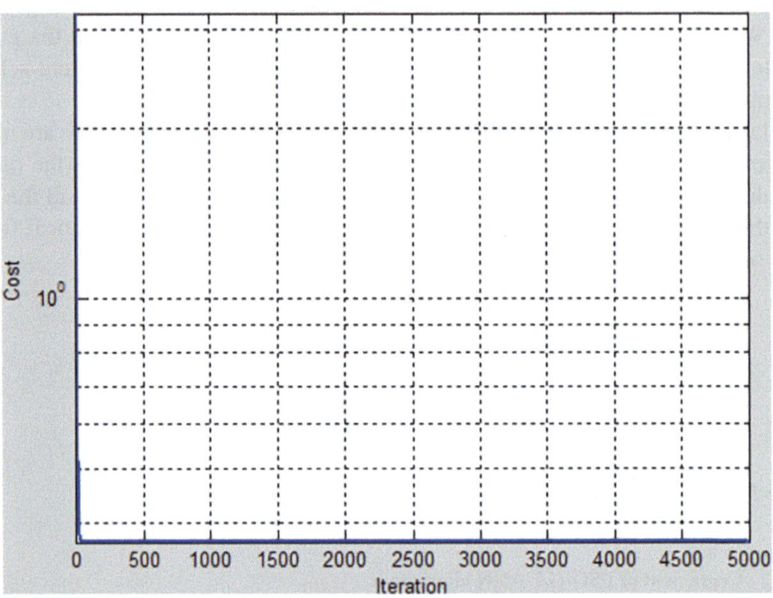

Fig. 2 Objective function of GA-LRS-ANN algorithm

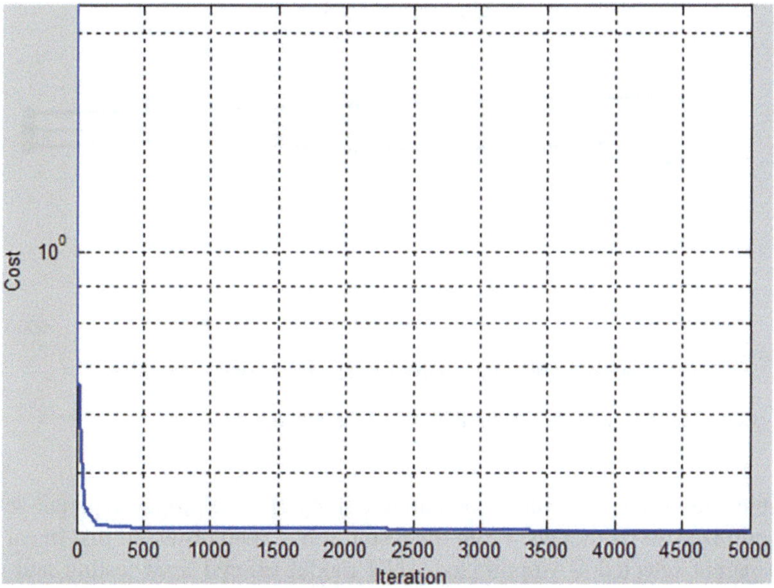

Fig. 3 Objective function of GA-TS-ANN algorithm

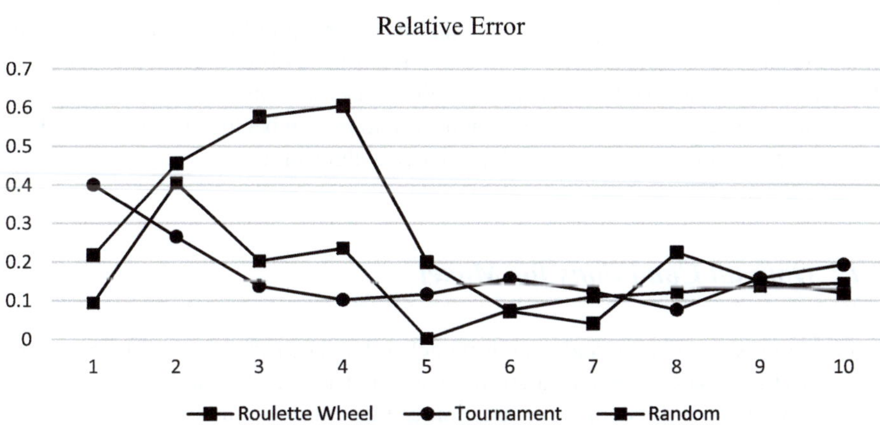

Fig. 4 Comparative study based on relative error

Table 3 Comparative based on different metrics

Name of the algorithm	Train RMSE	Test RMSE	Computational time in Sec	Average minimum fitness
GA-RWS-ANN	3.29577	7.2645	322.881934	0.003706
GA-LRS-ANN	1.85137	3.2345	318.566439	0.004519
GA-TS-ANN	1.95602	3.5698	515.098320	0.004139

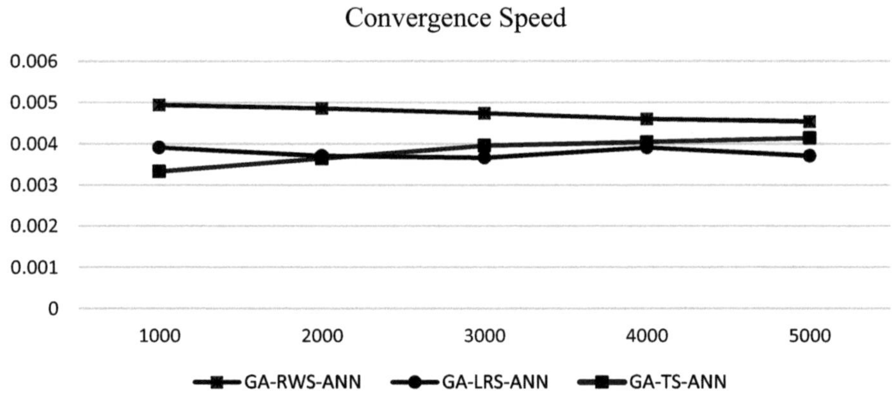

Fig. 5 Convergence speed of different PSO-GA-ANN algorithms

Table 3 represents an overall comparative study of three proposed parent-selected GA-optimized ANN models. From the result, it has been observed that linear rank-based parent-selected GA-optimized ANN model offered least testing and cross-validation root mean square error (RMSE), and least computational time but the Roulette wheel parent-based GA-optimized ANN model offered least average minimum fitness value.

An optimization technique can be called the best algorithm when it offers least fitness value over a certain time. From Fig. 5 it has been observed that the GA-LRS-ANN model reaches to the least fitness value very fast compared to the other two algorithms. Hence from cross cross-validation and testing perspective, the GA-LRS–ANN model is more superior than the other two algorithms.

3.1 Optimal Conditions for Model

From the cross-validation and testing model, it has been observed that linear rank selected (LRS) parent genetic algorithms-assisted artificial neural network is the best-fitted model and its objective function can be expressed as Eq. (13):

$$CSS = 4.1155 * CP + 0.2161 * SM + 3.515 * DB - 6.0187 \tag{13}$$

To obtain the optimized objective function for a given range of input it has been discovered that the best combination value for clay percentage by weight (CP), weighted geometric standard deviation of sediment mixture (SM), and dimensionless dry bulk unit weight of cohesive sediment (DB) are 0.104, 2.9523, and 1.6921 respectively to obtain critical shear stress of cohesive sediment, which gives quite satisfactory result of actual experimental data. To achieve an optimal condition of critical shear proposed algorithms take 28.365 sec as computational times with a

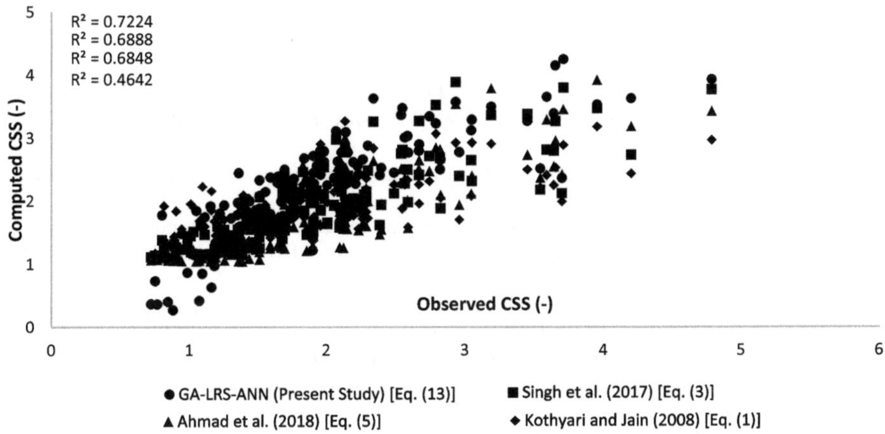

Fig. 6 Comparison of ANN-based model Eq. (13) with existing models of Eqs. (1), (3), and (5)

minimum fitness value of 2.356 E-05. The ANN proposed model as Eq. (13) compared with the existing models as Eqs. (1), (3), and (5) given by Kothyari and Jain (2008), Singh et al. (2017), and Ahmad et al. (2018) respectively is shown in Fig. 6. The ANN-based proposed model outperformed with better regression coefficient of 0.72 among all the models as depicted in Fig. 6.

4 Conclusions

The present study aims to design a mathematical model to compute critical shear stress of coarser particle in cohesive mixture of mobile channel bed and find out the optimum process parameters. For this purpose, ANN-optimized genetic algorithm was used with 148 experimental datasets from the present experimental study along with existing data from Kothyari and Jain (2008), Singh et al. (2017), and Ahmad et al. (2018). Three well-known selection operators GA-RWS-ANN, GA-LRS-ANN, and GA-TS-ANN were considered to optimize the ANN model, and their performance was analyzed and compared with the proposed model with existing ones. It has been found that the computed value obtained by GA-PSO-ANN closely matches with experimental result. The maximum absolute error for GA-ANN is 7.474% obtained by the GA-TS-ANN model during training datasets. Among all the three parent-selected GA optimized ANN models, GA-LRS-ANN model was outperformed. The proposed ANN model is represented by Eq. (13) and it outperformed when compared with the existing models given by Kothyari and Jain (2008), Singh et al. (2017), and Ahmad et al. (2018).

References

Ahmad, Z., Singh, U. K., & Kumar, A. (2018). Incipient motion for gravel particles in clay-silt-gravel cohesive mixtures. *Journal of Soils and Sediments, 18*(10), 3082–3093. https://doi.org/10.1007/s11368-017-1869-z

Garde, R. J., & RangaRaju, K. (2000). *Mechanics of sediment transportation and alluvial stream problems*. Taylor & Francis.

Kothyari, U. C., & Jain, R. K. (2008). Influence of cohesion on the incipient motion condition of sediment mixtures. *Water Resources Research, 44*, 1–15. https://doi.org/10.1029/2007WR006326

Rathore, D. S., Choudhary, A., & Agarwal, P. K. (2006). Assessment of sedimentation in Harakud Reservoir using digital remote sensing technique. *Journal of the Indian Society of Remote Sensing, 34*(Dec), 377–383. https://doi.org/10.1007/BF02990922

Shields, A. (1936). *Application of similarity principles and turbulence research to bed-load movement*. California Institute of Technology Pasadeza.

Singh, U. K., Ahmad, Z., Kumar, A., Pandey, M. (2019). Incipient motion for gravel particles in cohesionless sediment mixtures. *Iranian Journal of Science and Technology, Transactions of Civil Engineering, 43*(2), 253–262. https://doi.org/10.1007/s40996-018-0136-x.

Singh, U. K., & Ahmad, Z. (2019). Transport rate and bed profile computations for clay–silt–gravel mixture. *Environment and Earth Science, 78*, 432. https://doi.org/10.1007/s12665-019-8419-5

Singh, U. K., Ahmad, Z., & Kumar, A. (2017). Formulation for critical shear stress of cohesive sediment mixture. *Current Science, 113*, 2105–2111. https://doi.org/10.18520/cs/v113/i11/2105-2111

Singh, U. K., Jamei, M., Karbasi, M., Malik, A., and Pandey, M. (2022). Application of a modern multi-level ensemble approach for the estimation of critical shear stress in cohesive sediment mixture. *Journal of Hydrology, 127549.* https://doi.org/10.1016/j.jhydrol.2022.127549

An Integrative Approach for Oxygen Demand-Based Stream Water Quality Modelling Using QUAL2K-ANN Interactions

Chandrasekaran Sivapragasam, Ayingaran Ravinashree, and Mangottiri Vasudevan

Abstract The spatio-temporal variations in stream water quality are primarily due to the heat and mass exchanges occurring at the interfaces which are propagative and dilutive along the stretch. The streams are also susceptible for increased pollutant loading from various point and non-point sources causing an imminent depletion of dissolved oxygen (DO) concentration. Though water quality parameters are several, it is pertinent to identify the most critical factors for an easy, reliable and meaningful monitoring system. It is also necessary to have a simplified data-driven model for making necessary predictions for various decision-making processes. The present study investigates the characteristic relationship between various forms of oxygen demands (carbonaceous biochemical oxygen demand—cBOD and sediment oxygen demand—SOD) in a stream with the available DO for a selected stretch of River Bhavani in Tamil Nadu, India. The modelling framework consists of simulating a data-extensive water quality profile using QUAL2K and integrating the outputs for minimizing the required count of parameters using a multi-perceptron feed-forward artificial neural network (ANN) model. The proposed methodology suggested that a minimum of five parameters (inorganic suspended solids—ISS, DO, cBOD, SOD and total nitrogen—TN) are sufficient to simulate the concentration profiles with reasonable accuracy (R^2 varies from 0.88 to 0.95). For any local addition of organic-rich influent to the sediment load of the river, there is a corresponding change in the cBOD and DO within the selected stream profile. The results from the present study indicate the advantage of the combined modelling approach for prediction of water

C. Sivapragasam (✉) · A. Ravinashree
Department of Civil Engineering, Kalasalingam Academy of Research and Education, Krishnankoil, Tamil Nadu, India
e-mail: sivapragasam@klu.ac.in

M. Vasudevan (✉)
Department of Agricultural Engineering, Bannari Amman Institute of Technology, Sathyamangalam, Tamil Nadu, India
e-mail: vasudevan@bitsathy.ac.in

R. Srivastav, P. C. Nayak (eds.), *Applications of Machine Learning in Hydroclimatology*, The Springer Series in Applied Machine Learning,
https://doi.org/10.1007/978-3-031-64403-0_7

quality in case of complex interactions between various oxygen-demanding substrates.

1 Introduction

Water is one of the prime necessities of existence for all forms of life. The history of human civilization narrates the journey of exploration, utilization, consumption, exploitation, conservation, and reclamation of water resources at various time periods and places. Maintenance of water quality in rivers is of prime importance for every government to ensure the adequacy of public water supply and service infrastructure for the majority of the population. At least six out of the seventeen sustainable development goals (SDG) directly involve safe and responsible water management aspects. However, the majority of the surface water resources are extensively exploited not only by over-drafting and making changes in the ecological flows, but also by allowing various pollutants to enter and deteriorate the water quality to a large extent. The majority of rivers receiving an incessant flow of domestic or industrial effluents have almost lost much of their self-purification potential (Gopal & Sah, 1993; Das, 2011; Pratiwi et al., 2023). In order to minimize the adverse effects and to restore the water quality, the necessary managerial and legal interventions have to happen, which depends on the effective monitoring and modelling of water quality parameters from time to time.

The scope and efficiency of water quality models relies on their ability to reasonably depict the hydrodynamic routing as well as contaminant transport processes for a given simulation period across the designated flow paths. This involves simulation of various waste-loading scenarios by introducing the organic and inorganic pollutants to the streamflow and thereby predicting their spatio-temporal profile in conjunction with other environmental factors influencing the mass transfer mechanisms (Sivapragasam & Muttil, 2005; Gao & Li, 2015; Sankararajan et al., 2017; Chandrasekaran et al., 2017; Martin et al., 2018; Yuan et al., 2020). Typically, the contamination can be caused by either a point or nonpoint source entering a flowing freshwater stream. Though the point source of pollution (PS) can be easily located such as a pipe, dumpsite, or effluent outlet, the nonpoint source contaminants (NS) are difficult to locate back to their pathway and source of emission due to their invasive and diffusive nature.

Water quality models help us to understand the interactions between different components of a streamflow system as they are influenced by physical, chemical and biological features of the aqueous environment. Typical water quality models such as QUAL2K (Mangottiri et al., 2011; Mummidivarapu et al., 2023), MIKE (Nguyen et al., 2018; Badrzadeh et al., 2022), HEC-RAS (Mohammad et al., 2016; Muñoz et al., 2022), QUASAR, SWAT, and WASP have the provisions to incorporate the external pollutant sources across the stream stretch. Among them, the QUAL2K is an excellent model capable of simulating steady-state stream flow and continuous effluent-loading scenario for studying the waste load allocation entering a river from the nearby land area and in deriving suitable waste management strategies. The QUAL2K model is capable of representing the synergistic interaction between various sources of organic matter, micro-flora and fauna in the presence of various environmental and climatic factors. In other words, it ensures the significant responses of C-N-P cycles, algal-bacterial symbiosis, addition of benthic and carbonaceous demand, influence of air reaeration, and their combined effects on the dissolved oxygen dynamics.

Various researchers have employed QUAL2K and its different versions for simulation of waste load allocation scenarios and other planning exercises, particularly under the conditions of low -flow streams carrying constant and continuous effluent release conditions. Mangottiri et al. (2011) used the QUAL2E version of the model to simulate various waste-loading scenarios for the River Yamuna near New Delhi, India based on the carbon-nitrogen-oxygen cycles available with the model. Sarda (2013) applied QUAL2K for water quality modelling for a 15 km stretch of Godavari River in India and conducted a sensitivity analysis to identify the criticality of the parameters. They found that the model predictions were matching with the field data with some exceptions and the results are quite sensitive to the input parameters such as total nitrogen (TN), carbonaceous biochemical oxygen demand (cBOD) and depth coefficient. Gupta et al. (2013) observed a maximum of 10% deviation in the BOD data after including the reaeration model. Zhang et al. (2012) conducted a study on the Taihu Lake and Hongqui River in China for three scenarios having different water treatment technologies with varying configurations. They achieved reduction rates for BOD, TN, ammonia-nitrogen (NH_3–N), and total phosphorus (TP) as 49.50%, 32.81%, 35.94%, and 45.27%, respectively. Later, Idris et al. (2016) studied the surface water quality of Yamuna River in India using the QUAL2K model using various point source inputs. They reported that the model is highly sensitive to hydraulic parameters corresponding to the PS discharges and moderately sensitive to the degradation rate of fast cBOD and organic nitrogen. A stretch of 12.63 km of the Pamba River in India was studied by Ashwani et al. (2017) for the variations in water quality parameters (BOD, TSS, TN, TP, and alkalinity) using QUAL2K and the results obtained were well in agreement with the observed data.

Among the various pathways of organic matter degradation system addressed by various models, the predicament of sediment oxygen demand (SOD) is rarely investigated in detail including its direct causes and consequences. The nature of water flow and nutrient cycling within the riverbed may favour the benthic organisms to flourish and the corresponding oxygen demand can potentially affect the overall water quality. Since the source and mechanisms are not the same as the normal BOD, the model has to consider the SOD as a separate entity. The QUAL2K model enables us to study the supplementary action of SOD for the degradation of organic matter under the influence of temperature, sunshine penetration, growing rate of phytoplankton and zooplankton (Chapra et al., 2003; Ye et al., 2013).

Owing to the lack of sufficient validating database and due to the variability existing in the confining assumptions in the model, none of the models can be confirmed to be stand-alone for predicting the water quality scenario. Typically,

QUAL2K and WASP models are highly data-intensive and involve complicated mass transfer mechanisms thereby adding more sensitivity issues to the model. Therefore, it behoves us to prefer integrative approach for the development of hybrid models that can compensate the performance criteria of individual modelling components resulting in a better prediction of the intended output parameters. In an earlier attempt, Fan et al. (2009) attempted to simulate the tidal effects on river water quality by integrating QUAL2K with HEC-RAS with a combination of C-N-P cycling scenario. Considering more of a hydrologic problem, the impacts of later inflows were successfully modelled by integrating SWAT with QUAL2K by Bui et al. (2019). Later, Kamal et al. (2020) developed QUAL2K-GIS integration for simulating pollutant-loading scenario. The HEC-RAS in combination with QUAL2K was found to be successful in assessing the pollutant-carrying capacity of the river (Fan et al., 2021). Recently, Mummidivarapu et al. (2023) developed an approach for the integration of QUAL2K with Global Environmental Flow Calculator (GEFC) and Geographical Information System (GIS) to develop water quality index (WQI)-based map charts of water quality.

The use of artificial neural network (ANN) models in solving data-intensive problems has been quite successful in environmental and water resources engineering applications (Bowden et al., 2002; Palani et al., 2008; Sarkar & Pandey, 2015; Mahalakshmi et al., 2019; Sharma et al., 2019; Sivapragasam et al., 2023). With the advent of machine learning algorithms, more sophisticated models are emerging to make simplified computations with faster and reliable outputs. As the core of water quality assessment necessarily deals with DO-BOD kinetics, most of the ANN-based models are developed for predicting the DO for the given set of input BOD values. Ahmed (2017) developed a feed-forward and radially functional ANN model to predict DO based on input BOD values with reasonable accuracy ($R > 0.90$ and $\text{MSE} = 0.5$). Similarly, Niroobakhsh et al. (2012) used a multi-layer perceptron (MLP) for predicting the total dissolved solid (TDS) concentrations for the Jajrood River of Iran with more than 90% accuracy.

Although ANN-based models are quite capable of smart predictions, they often lack accuracy due to the lack of reliable input data for training, testing, and validation. In exception of the measured parameters of water quality parameters, QUAL2K offers reasonably reliable simulation of water quality for river stretches, thereby enabling different case studies to generate adequate dataset for further predictive analysis. The need of the present study can be identified as the scarcity of suitable integrative approach for exploring the simulation capabilities of QUAL2K towards evaluating the performance of an ANN-based predictive model. Further, as mentioned earlier, the studies related to the role of SOD in influencing the DO-BOD kinetics during the organic as well as inorganic suspended solids (ISS) during various waste-loading scenarios are very limited. Thus, the objective of the study is to develop an integrative framework for simulating the SOD profile in the Bhavani River of Tamil Nadu, India for various combinations of DO, cBOD and ISS concentrations and to derive an optimized number of input parameters for predicting the SOD concentration using a back propagation-based ANN model.

2 Materials and Methods

2.1 *Study Area*

The proposed modelling framework was adopted for a selected portion of a linear stretch of the Bhavani River in the Erode district of Tamil Nadu, India. The study area comes under the Lower Bhavani Irrigation Project covering a total stretch of 251 km of channel-based irrigation system. It is observed that the surface water utilization can be maximized in a closed-type layout of the basin, if sufficiently harvested. However, there is a typical concern of drawback about this layout on the water quality due to the persistence of pollutants emerging from the anthropogenic activities within the study area as well as from the neighbouring industrial belt. Therefore, the present study also aims to provide a sensible characterization of various developmental strategies for simulating futuristic water quality under limited datasets.

2.2 *Setting Up of QUAL2K Model*

On a broader aspect, the pollutant concentrations are highly variant according to the hydraulic and meteorological conditions of the river during the study period. In order to represent wider distribution of water quality and associated flow properties, a length of 100 km stretch of the river was delineated. The total river stretch is divided into 54 reaches (small portions) with an assumed headwater flow rate of 2.0 m^3/s at steady-state conditions. Further, to represent the deviations in water quality along the stretch, a total of 5 point sources of pollution were added at selected intervals indicating the aggregate effect of effluent discharge. In addition, a non-point source was also considered to compare the pollutant (or nutrient) contributions originating from the farmlands and industrial zones (Ravinashree et al., 2022).

Additional input dataset for the spreadsheet-based QUAL2K model consists of influent water quality parameters such as pH, temperature, ISS, DO, cBOD and phytoplankton concentration. A combination of physical, chemical and biological parameters was selected for this study under the prelude of simulating the required scenario of sediment-driven biotic environment. In other words, the scenario implies that most of the pollution originates from a variety of sources, and the combined effect can be experienced all at once at a downstream location. The point sources of pollution were denoted at the river reaches corresponding to the distance from the headwater as 12.17 km, 27.22 km, 42.14 km, 57.14 km, and 97.15 km. Most of the required input data were based on the available information collected from recent literature as well as based on the typical assumption for the standard parameters (including reaction rate coefficients) as shown in Table 1 (Ravinashree et al., 2022). The influence of diffuse source is simulated by varying the inflow rate (0.1, and 0.5 m^3/s) as well as by varying the abstraction rate (0.1, and 0.5 m^3/s). The

Table 1 Summary of input parameters used for the QUAL2K model for River Bhavani

Parameter	Value	Unit	Remarks
Headwater flow rate	2.0	m^3/s	Taken as the average value—Varies between 0.1 and 10 m^3/s for calibration
Manning's constant (n)	0.045	–	Suitable for the channel with some weeds and stones
Temperature	27	°C	Varies diurnally and seasonally
Conductivity	800	umhos	Presence of dissolved ions
Inorganic suspended solids	50	mg/L	Presence of sediment and mineral-rich soil
Dissolved oxygen	5	mg/L	Varies between 4 and 8 depending on pollution
$CBOD_{slow}$	10	mg O_2/L	Slightly complex organic matter
$CBOD_{fast}$	5	mg O_2/L	Readily degradable organic matter
Organic nitrogen	1800	ug/L	Indicative of agricultural pollution
Ammoniac nitrogen	500	ug/L	Indicative of undigested organic matter
Nitrate-nitrogen	300	ug/L	Indicative of depletion of organic matter
Organic phosphorus	10	ug/L	Indicative of agricultural pollution
Inorganic phosphorus (SRP)	20	ug/L	Indicative of sediment and mineral-rich soil
Phytoplankton	50	ug/L	Representative of nutrient absorption and photosynthesis
Internal nitrogen (INP)	500	ug/L	Indicative of significant microbial transformations
Internal phosphorus (IPP)	30	ug/L	Indicative of significant microbial transformations
Detritus (POM)	500	mg D/L	Indicative of organic debris present in water
Pathogens	30	cfu/100 mL	Presence of total coliforms
Alkalinity	150	mg $CaCO_3$/L	Indicative of the buffering capacity of water
pH	8.0	–	Indicative of overall chemical stability of water

introduction of the point sources was represented by the assigned values of inflow rate (0.05, and 0.1 m^3/s) at the selected locations as shown in Fig. 1.

2.3 *Calibration and Sensitivity Analysis of the QUAL2K Model*

As the primary goal of this study is to forecast the water quality characteristics of the stream, it is required to calibrate the model initially for the flow simulation component. For this, three conditions were considered as possible means of accessing

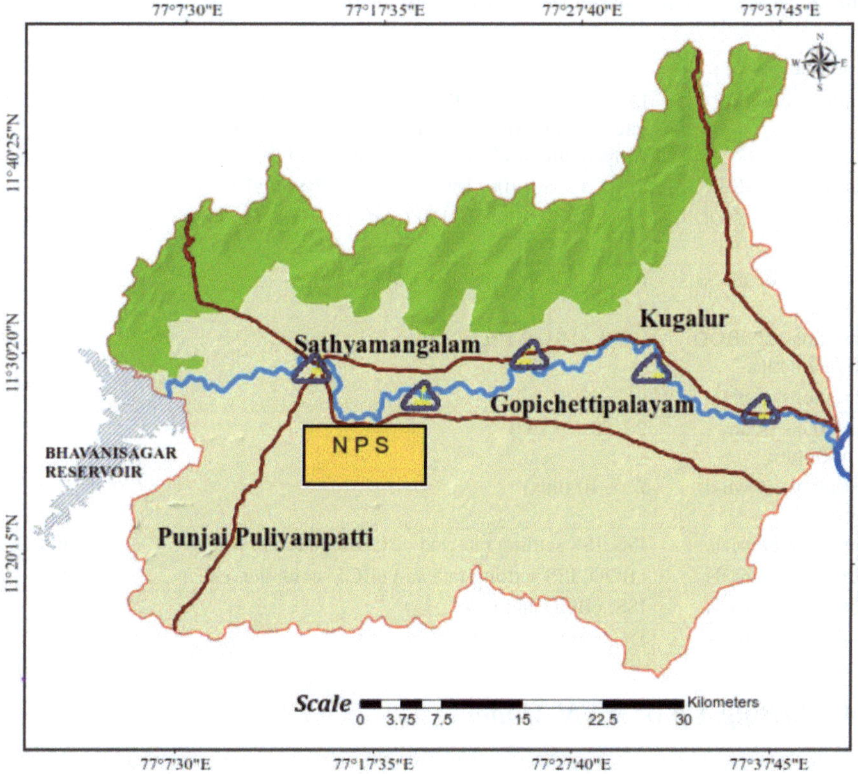

Fig. 1 Representation of the study area showing the selected stretch of the river with the locations of point source (marked as triangular nodes) and non-point source (marked as NPS) of pollution

realistic discharge values—Manning's formula, Weir formula, and Rating curve. For each case, the model is run to fine-tune the corresponding flow model parameters (such as velocity coefficient, depth coefficient and roughness coefficient), while other parameters were kept unaltered during each run in order to check the satisfactory match with the reported results.

The QUAL2K model is further analysed for the sensitivity of various modelling parameters such as discretization limit, headwater quality parameters and pollutant loading rates. A series of case studies were framed for conducting the simulation experiments by varying the limiting values under the low, medium and high categories. This approach has not only helped to identify the independent sensitivity towards individual parameters, but also to estimate the extent to which the results (i.e. downstream water quality profile) can vary for a given degree of deviations in the input parameters. In addition, various combinations of the input parameters were selected for the case studies to further understand their combinatorial impacts on the evolution of water quality throughout the river stretch. The details of the case study are represented in Table 2.

Table 2 Selected parameters and their variations for different case studies

Case studies	Parameters	No. of trials
Control condition	ISS = 50 mg/L; cBOD = 10 mg/L; DO = 4 mg/L; ISS settling rate = 0.05; cBOD oxidation rate = 0.01; pH = 7; air temperature = 25 °C; wind speed = 5 m/s	1
Variations of ISS	5, 10, 15, 50, 100, 150, 250, 500, 750 (mg/L)	6
Variations of cBOD	5, 10, 15, 20, 30, 50, 100 (mg/L)	7
Variations of DO	4, 7, 10 (mg/L)	3
Variations of ISS settling rate	0.005, 0.05, 0.5 (day^{-1})	3
Variations of cBOD oxidation rate	0.001, 0.01, 0.1 (day^{-1})	3
Variations of pH	4, 7, 10	3
Variations of air temperature	20, 25, 30 (°C)	3
Variations of wind speed	2, 5, 10 (m/s)	3
Combination trials	ISS, ISS settling rate and cBOD oxidation rate	6
Combination trials	cBOD, ISS settling rate and cBOD oxidation rate	6
Combination trials	ISS, cBOD and DO	6
Combination trials	DO, cBOD and pH	6

2.4 *Setting Up of ANN Model*

The ANN model is setup for the study using the Neuroshell 2® (Release 4.0) environment by selecting suitable characteristic model parameters (Antanasijević et al., 2013). In order to enable the smooth convergence of the ANN model, the input parameters are separated as training set, testing set and validation set by subsequently extracting and importing them as per the preferred proportion. The ANN model is allowed to undergo sufficient learning using the training dataset from the available output and further optimizing the architecture (in terms of the design of layers and selection of neurons at each layer). For this, a suitable number of trials (mentioned as case studies above) were selected from the QUAL2K results corresponding to the variations in the input parameters (ISS, cBOD, DO, pH, phytoplankton, TP and nitrogen forms) at three locations (0 km, 12.17 km and 27.22 km). As mentioned above, the SOD concentration as the preferred output is validated by using a three-layered propagation network for the training. Out of the total 56 case studies, results from 36 case studies were used for training the model, where the outputs from 10 each of the remaining case studies were employed for testing and validating the model. A two-phase optimization of the architecture was performed by varying the number of neurons, and further, by reducing the number of input parameters to the ANN model.

2.5 Model Integration Protocol

Based on the sensitivity analysis, it is realized that although QUAL2K is capable of producing sufficiently large database based on the possible complex interactions to result in using the available input data as much as 35–40 parameters, there are actually only a few critical parameters which can significantly influence the required output - sediment oxygen demand (SOD). With this confirmation, the proposed integrated modelling framework utilizes the fine-tuned results from the basic flow model (QUAL2K) as input for running the predictive ANN architecture. This is implemented by converting the outputs from the QUAL2K model into ASCII format and feeding to the ANN model, thereby maintaining the input-output coordination intact after every model run. The overall methodology adopted in the present study for the prediction of SOD through two different model platforms can be summarized as shown in Fig. 2.

3 Results and Discussion

3.1 Spatial Distribution of Inorganic Pollutants

The hydraulic and environmental features of the selected stretch of River Bhavani were simulated using the proposed modelling approach based on the conceptual model and their implicit assumptions. As the initial step, the QUAL2K model was calibrated using the flow model (taken as the control condition) and the corresponding quality model was further verified for the contaminant parameters mentioned under the control conditions. We have observed that the basic river flow profile for three scenarios was consistently predicted for the selected reaches of the river (Ravinashree et al., 2022).

The inorganic pollutants as released through the constant continuous point sources from the specified locations were found to exhibit an increasing effect to reach the highest concentration at the outlet. However, based on the initial concentrations and settling rate of the ISS, the concentration values were found to decrease in the middle reaches before getting the cumulative effect towards the end (Fig. 3). The decrease in ISS concentration over the selected reach points (which are close to the entry of the PS locations) is found to be marginal for smaller initial values of ISS. However, the higher ISS input has resulted in a substantially high stable concentration throughout the river stretch, indicating the significance of resuspension and deposition during the turbid water flow (Chapra et al., 2003). This may give rise to two scenarios: either the solid particles in the influent are quite discrete and may easily deposit to the river bed as they enter with the specified low discharge values, or the shallow depth of the river may provide high mixing and flocculent settling during the travel. In either case, a cumulative effect of inorganic pollutants can be expected to contribute towards the river bed sediments. Therefore, simulating the

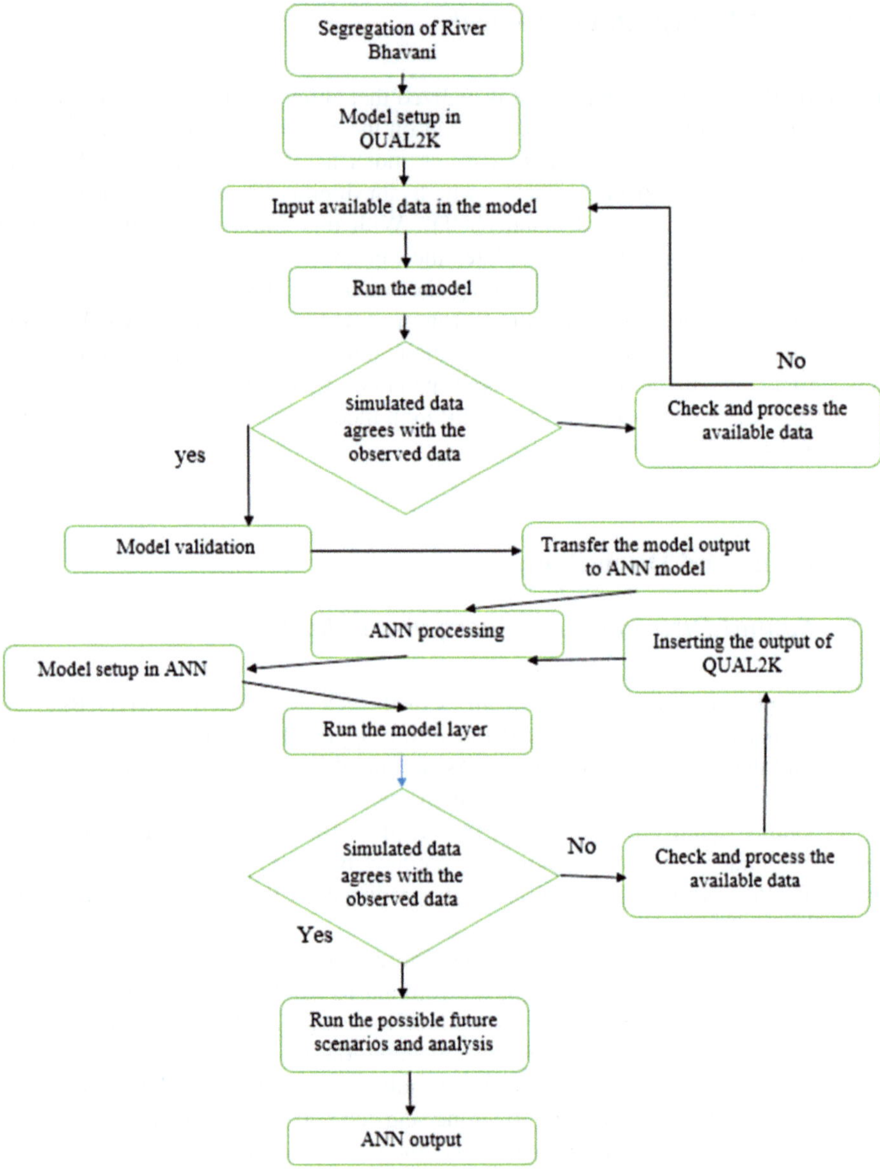

Fig. 2 The overall methodology of the modelling framework

concentration profile of ISS can reasonably explain the nature and extent of sediment trapping at the river bed.

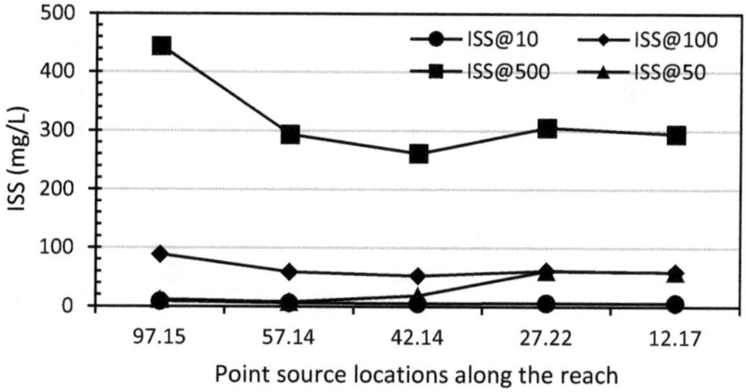

Fig. 3 Variations in the concentration of ISS at selected reach points (mentioned in km). Each curve represents a unique combination of input parameters while only one parameter (ISS) is highlighted

3.2 *Spatial Distribution of Organic Pollutants*

The extent of organic pollution in the river is evaluated using several simulation case studies in terms of fluctuations in DO and cBOD. It is to be noted that the QUAL2K model is capable of incorporating the diurnal and seasonal fluctuations in water temperature, thereby, the DO concentration is considered to represent a varying profile for the selected stretch of the river. Now, the simulation of organic pollutants in QUAL2K mainly considers two aspects: carbonaceous oxygen demand and nitrogenous oxygen demand. Considering the scenario of unrestricted disposal of domestic effluents through several identifiable point sources, the prevailing biochemical environment will be most suitable for quicker degradation of simple organic compounds by making use of the available dissolved oxygen and microfauna present in the river (Zhang et al., 2012). Therefore, an assimilative mechanism can be easily functional within the immediate downstream side of the point sources that will further undergo modification (mostly from the carbonaceous demand to nitrogenous demand) during the travel.

The simulation results indicate that the contamination of cBOD is drastically increasing within the chosen flow regime despite the fact that DO is adequately supplied throughout the flow domain (Fig. 4). In other words, there is no significant impact of the initial DO concentration on the cBOD profile which is an indicative of the cumulative nature of organic matter in the water body. This is an indication of the net release of organic contaminants due to the continuous waste-loading conditions (Antanasijević et al., 2013; Chandrasekaran et al., 2017). It is to be noted that the impact of all assimilative reactions happening within the river path was either instantaneous or locally effective, thereby demonstrating a net increase in the contaminant concentration. As confirmed from the literature, the existence of adequate amount of DO is a direct evidence of controlled organic pollution, which in fact, alleviates many intrinsic subsidiary reactions. Therefore, higher values of

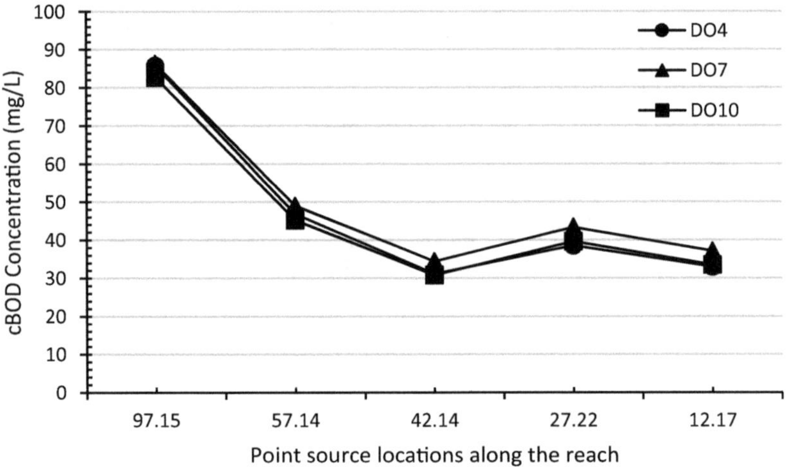

Fig. 4 Variations in the concentration of cBOD at selected reach points (mentioned in km). Each curve represents a unique combination of input parameters while only one parameter (DO) is highlighted

cBOD are of concern in order to give an adequate degradation period because they are expected to consist of fast-degrading species.

As mentioned above, the corresponding trend of the DO profile can provide further details on the oxygen-limited mechanisms prevailing in the river stretch. The calibrated QUAL2K model is capable of simulating the DO profile which showed a gradual increase and decrease before and after the middle stretch of the river (Fig. 5). This confirms that following the entry of pollutants from the selected point sources, the depletion of DO was very quick despite the regenerative mechanism due to the mass transfer equilibrium with the atmospheric air (Ashwani et al., 2017; Mahalakshmi et al., 2019). It is to be noted that each curve in Fig. 5 represents unique combinations of ISS, DO, cBOD and pH as part of the case studies mentioned in Table 2 and is to be understood with reference to the combination of input parameters taken for each case study. The highest depletion in the DO profile was attributed to the scenario when the initial concentration of DO, cBOD, phytoplankton and ISS were 10 mg/L, 30 mg/L, 15 mg/L and 250 mg/L respectively. This is probably due to the increased biochemical activity in the presence of the phytoplankton present in the river. This is also in agreement with the reduced concentration profile of cBOD loading as shown in Fig. 4 (Ravinashree et al., 2022). Similarly, when the initial cBOD was quite high (i.e. 100 mg/L), the depletion of DO was lowest with low initial DO (4 mg/L) and low cBOD (10 mg/L) despite the presence of other factors. It is also observed that the river has good potential for re-oxygenation to replenish the depleted DO within the intervals of point loads of pollutants. Another important aspect of DO consumption in the river is supposed to be contributed from the sediment oxygen demand (SOD) which bears a reciprocal

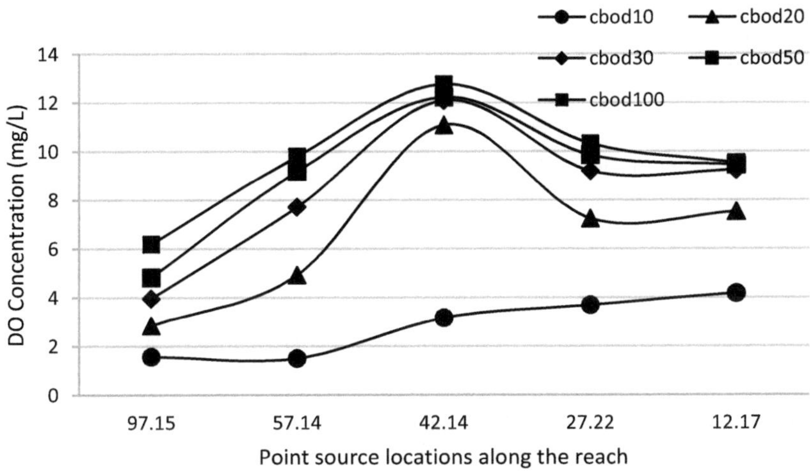

Fig. 5 Variations in the concentration of DO at selected reach points (mentioned in km). Each curve represents a unique combination of input parameters while only one parameter (cBOD) is highlighted

relationship with DO. This is further explained in detail with the help of the integrated Q2K-ANN model framework proposed in this study.

3.3 *Variations in Sediment Oxygen Demand*

The spatial variation of sediment oxygen demand (SOD) has been emphatically addressed to understand its contribution towards the prevailing cBOD-DO interactions in the aqueous environment. As evident from the literature, the sediment is preferably considered as a dominant oxygen sink in surface water quality assessment and is critical towards maintaining the biotic ecosystem in balance (Mangottiri et al., 2011; Liu & Chen, 2012; Mohammad et al., 2016). The depletion of dissolved oxygen due to the induced biochemical reactions from the sediments is, therefore, attributed to an additional organic load on the mass transfer and transport mechanism of DO within the river path. The results from the QUAL2K simulation indicate that the SOD profile remains more or less uniform within the selected stretch of the river, while it varies significantly for the selected combination of model parameters based on their initial concentrations (Fig. 6). The SOD for the scenario ISS500 represents the lowest concentration typically indicating the predominance of inorganic pollutants in the river (ISS = 500 mg/L, cBOD = 10 mg/L and DO = 5 mg/L) especially during the turbulent flow carrying turbid water. The second scenario (PHY30) indicates the presence of phytoplankton in the water body that can influence the photosensitive mass transfer mechanisms between the sediment, micro-organisms and algal species present in the water. In contradiction to the previous scenario, the

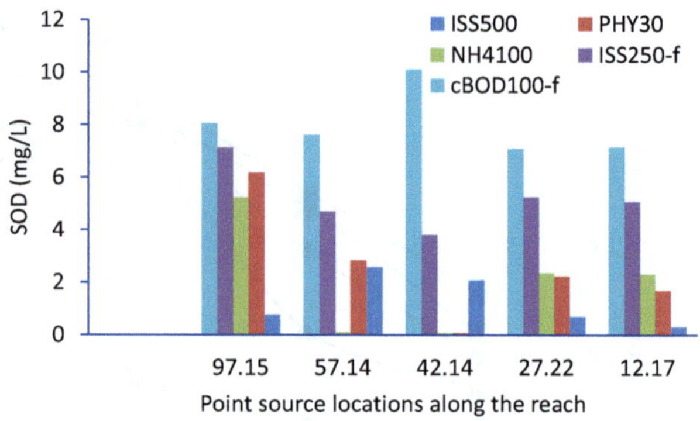

Fig. 6 Variations in the concentration of SOD at selected reach points (mentioned in km). Each bar represents a unique combination of input parameters while only a few parameters are highlighted

amount of SOD during the middle reaches was extremely low in the presence of phytoplankton compared to the presence of ISS, which is also accompanied by high DO and low cBOD concentrations. The key factor, nonetheless, is DO, which is responsible for controlling the SOD and cBOD such that a synergistic algal-bacterial symbiosis can be achieved for a significant portion of the river.

The third scenario (NH4100) has similar inorganic contribution but additionally contributes high concentration of ammonia-nitrogen (NH_4-N), thereby increasing the oxygen demand especially towards the outlet reaches. In other words, the increase in SOD for the high NH_4-N scenario also indicates that preferable utilization of cBOD has been accomplished before consuming the nitrogenous organic matter. This is a very important inference to understand how the SOD can directly influence the existing cBOD-DO interactions. We have also considered two scenarios of quick reactions for inorganic (ISS settling rate of 0.5/day) and organic (cBOD oxidation rate 0.1/day) pollution conditions. When the initial concentration and settling rate of ISS are high (ISS250-f), the SOD concentration was persistent throughout the river stretch though a marginal reduction was observed during the middle reaches. However, the highest concentration of SOD was obtained when the cBOD settling rate was considered along with a high initial concentration. This scenario has resulted in creating multiple oxygen demands simultaneously and peak was observed during the middle reaches, indicating the limiting scenario. It may be concluded from the results that when there is a minimum external loading to the river as observed in some of the above scenarios, the SOD can be attributed as the only cause for the change in DO compared to the cBOD sink.

Apart from the scenario-based approach for multiple parameters influencing SOD at a time, their individual correlations were also analysed in the study. For this, we have selected the evolution of SOD concentration at a location 16 km away from the starting point and the variations in four most critical parameters (ISS, pH, DO, and cBOD) were observed along with SOD values (Fig. 7). It is observed that ISS and DO show inverse relationship with SOD while cBOD is in direction correlation with SOD. It is obvious to realize that depletion or regeneration of SOD due to the organic portion of the contaminants is directly influenced by DO, which is reflected in the response with cBOD, whereas an increase in ISS can hamper the photosensitive

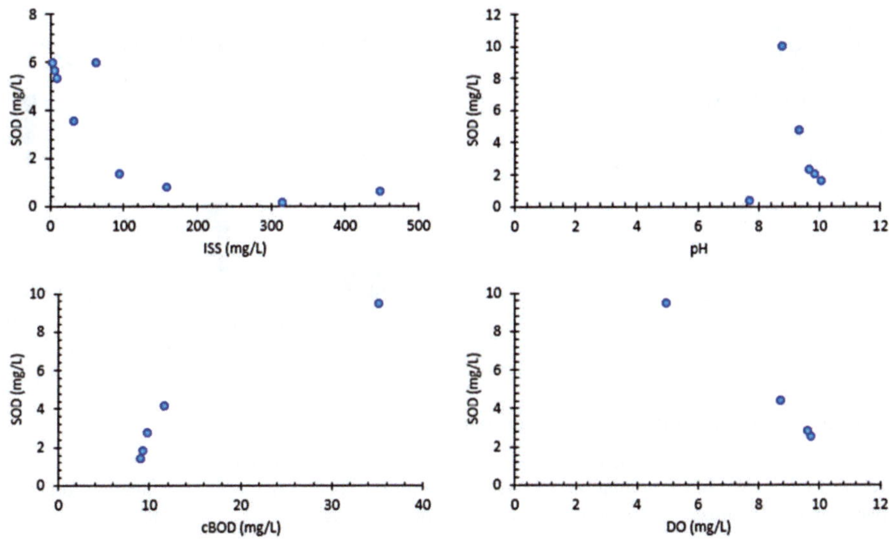

Fig. 7 Variations in the concentration of SOD as influenced by variations in ISS, pH, cBOD and DO at a point located at 16 km on the river stretch

biochemical reactions (Bilotta & Brazier, 2008; Sharma et al., 2009). However, there is no significant influence of pH on the SOD concentrations, which may be partly due to the fact that the redox conditions are not much influential on the physical settling of the pollutants, which in turn contributes towards the SOD. The progressive trend indicates that SOD and cBOD have similar trends of spatial variations corresponding to the variations in the DO levels at the selected reaches. This is attributed to the combined effect of oxygen consumption and retrieval experienced by the river during the reception of continuous flow from the selected PS locations. Based on the results, it may be inferred that the self-purifying capacity of the river is significantly affected by the inherent concentrations of SOD, which is in agreement with the presence of ISS along the same locations.

3.4 ANN-Based Prediction of Sediment Oxygen Demand

The datasets derived from the preliminary independent modelling analyses (using QUAL2K) were used to obtain the hybrid model. The integration with the ANN model is performed for predicting the concentration of SOD as the preferred output against a pre-defined set of ten input parameters (ISS, DO, cBOD, NH_4, NO_3, PHY, pH, TN, TP, and TKN). All these parameters are obtained as the outcome of QUAL2K simulation, which were analysed and compared under various scenarios for testing their consistency. Based on the sensitivity analysis, the input parameters were chosen so as to efficiently minimize the modelling errors. Through this

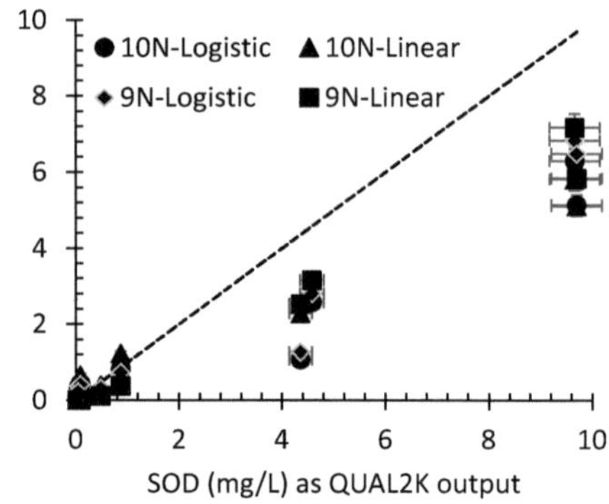

Fig. 8 Comparison of predicted SOD (mg/L) by ANN model with corresponding QUAL2K model based on 10 input parameters

Table 3 A summary of the network optimization analysis

No. of input parameters	No. of neurons	Type of activation function	Coefficient of correlation (R)	Coefficient of determination (R^2)	Root mean squared error (RMSE)
10	10	Logistic	0.965	0.932	2.172
10	10	Linear	0.986	0.972	2.070
10	9	Logistic	0.974	0.948	1.768
10	9	Linear	0.992	0.983	3.195
5	10	Logistic	0.956	0.915	2.179
5	10	Linear	0.976	0.915	1.232
5	9	Logistic	0.949	0.901	2.122
5	9	Linear	0.931	0.867	2.518

sampling strategy, the prediction errors were minimized during the training and testing stages. There were 5 training and 2 testing patterns during the learning phase, which converged to a least average error of $1.25\text{x}10^{-4}$ within 2,76,300 epochs consuming about 82 seconds of computational time. A number of trial runs were performed to obtain best validation results based on the consistency during the training and testing phases (Fig. 8). The model was optimized for the number of neurons (9 and 10) and the type of activation function (linear and logistic) for the analysis. It was observed that the linear function has predicted the SOD concentration with highest correlation and with minimum residual errors when the number of neurons was 10 (Table 3). Further, when the number of neurons was reduced to 9, the coefficient of correlation increased but with an increase in the error values.

The Q2K model has invariably resulted in some amount of training error compared to the testing error due to the fact that the training dataset encompasses virtually all conceivable scenarios using the 10 input parameters (Ahmed, 2017; Wai et al., 2022). However, it is envisaged that all the ten parameters may not have significant influence on the SOD concentration as the output. Therefore, in an attempt to minimize the simulation effort, the number of input parameters is reduced to merely 5 (ISS, DO, cBOD, pH, and PHYTO) and the neural network simulations were repeated for convergence. The results show that there is an apparent

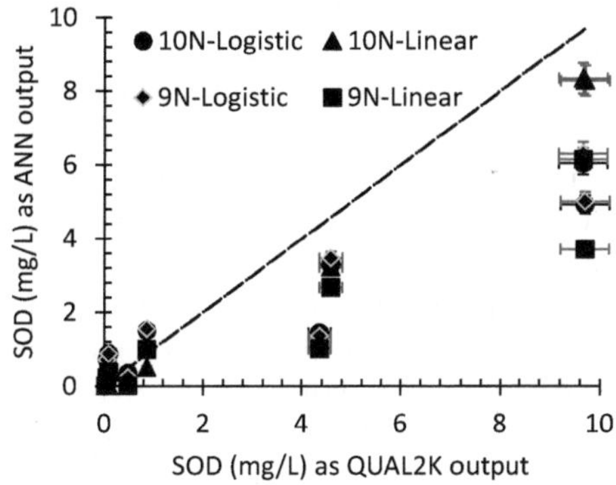

Fig. 9 Comparison of predicted SOD (mg/L) by ANN model with corresponding QUAL2K model based on 5 input parameters

improvement in the correlation values by removing some of the less influential parameters from the input dataset, thus making the model more compatible (Fig. 9). Though there is a slight reduction in the correlation coefficients, the numerical values of the errors for the 5-parameter model were lower than those for the 10-parameter model. Both of the exercises confirm that the ANN model with a linear function is better suitable for the better prediction of SOD than the logistic function. A summary of the network optimization analysis is provided in Table 3.

Based on this integrated modelling exercise, it is evident that all the case studies mentioned in the QUAL2K modelling are not necessary to provide the required output from the ANN model. The QUAL2K, being a highly parameter-centric model, can provide a large number of combinations for simulating various hypothetical scenarios; however, the most significant scenarios alone are sufficient for the ANN model to predict the output SOD concentration. The ANN being the key element of the proposed data-driven model is found to be well-trained for the case studies with reduced number of parameters without significantly losing the prediction accuracy and consistency. It may be inferred that although QUAL2K is necessary to provide the benchmark conditions of the contamination scenario, the futuristic prediction of the output concentration using the ANN model can be achieved with a lesser number of parameters, without demanding hypothetical assumptions in the QUAL2K model. Hence the proposed modelling framework can be effectively utilized for simulating and predicting the river water quality scenario in case of limited observed data.

4 Conclusion

The present study aims at developing an integrated modelling framework for simulating the river water quality by making advantage of the data-enrichment capability of QUAL2K model and self-learning ability of ANN model. The water quality of River Bhavani in Tamil Nadu, India is studied for a stretch of 100 km by considering suitable headwater and pollutant source conditions. The results indicate that QUAL2K model was capable of generating a large dataset for the water quality based on the supplied input data which were suitably integrated to the ANN model for prediction of sediment oxygen demand. The combined modelling approach revealed that irrespective of the nature of origin (organic or inorganic), the concentration of sediments has a significant role in limiting the available oxygen in the river water. There exists a non-linear correlation between the bCOD and ISS towards SOD individually which is better predicted for higher input concentrations, indicating the significance of external pollutant sources. The results indicate that the regenerative capacity of River Bhavani has been critically affected by the organic-rich effluents from the point sources which demands managerial interventions to reduce the waste inflow to the river. In this regard, the complementary nature of sharing the available dissolved oxygen between the persisted sediments and the externally introduced pollutant loads (organic and inorganic type) requires special attention for simulating the mass transfer dynamics in a perennial stream. The proposed modelling framework provides adequate accuracy in prediction by using a reduced parameter model for the sediment oxygen demand. Such modelling studies are required to be further optimized for simulating the combinatorial and compensative interactions between various water quality parameters.

References

Ahmed, A. M. (2017). Prediction of dissolved oxygen in Surma River by biochemical oxygen demand and chemical oxygen demand using the artificial neural networks (ANNs). *Journal of King Saud University-Engineering Sciences, 29*(2), 151–158. https://doi.org/10.1016/j.jksues.2014.05.001

Antanasijević, D., Pocajt, V., Povrenović, D., Perić-Grujić, A., & Ristić, M. (2013). Modelling of dissolved oxygen content using artificial neural networks: Danube River, North Serbia, case study. *Environmental Science and Pollution Research, 20*, 9006–9013. https://doi.org/10.1007/s11356-013-1876-6

Ashwani, S., Vivek, B., Ratnoji, S., Jayakumar, P., & Jainet, P. J. (2017). Application of QUAL2K model for prediction of water quality in a selected stretch of Pamba River. *International Journal of Civil Engineering and Technology, 8*(6), 75–84. https://iaeme.com/Home/article_id/IJCIET_08_06_009

Badrzadeh, N., Samani, J. M. V., Mazaheri, M., & Kuriqi, A. (2022). Evaluation of management practices on agricultural nonpoint source pollution discharges into the rivers under climate change effects. *Science of the Total Environment, 838*, 156643. https://doi.org/10.1016/j.scitotenv.2022.156643

Bilotta, G. S., & Brazier, R. E. (2008). Understanding the influence of suspended solids on water quality and aquatic biota. *Water research, 42*(12), 2849–2861.

Bowden, G. J., Maier, H. R., & Dandy, G. C. (2002). Optimal division of data for neural network models in water resources applications. *Water Resources Research, 38*(2), 2–1. https://doi.org/10.1029/2001WR000266

Bui, H. H., Ha, N. H., Nguyen, T. N. D., Nguyen, A. T., Pham, T. T. H., Kandasamy, J., & Nguyen, T. V. (2019). Integration of SWAT and QUAL2K for water quality modeling in a data scarce basin of Cau River basin in Vietnam. *Ecohydrology & Hydrobiology, 19*(2), 210–223. https://doi.org/10.1016/j.ecohyd.2019.03.005

Chandrasekaran, S., Sankararajan, V., Neelakandhan, N., & Ram Kumar, M. (2017). Genetic programming-based mathematical modeling of influence of weather parameters in BOD 5 removal by Lemna minor. *Environmental Monitoring and Assessment, 189*, 1–12. https://doi.org/10.1007/s10661-017-6337-0

Chapra, S. C., Pelletier, G. J., & Tao, H. (2003). Qual2K. *A modeling framework for simulating river and stream water quality (beta version): Documentation and user's manual, Civil and Environmental Engineering Department, Medford, Tufts University*. https://csdms.colorado.edu/csdms_wiki/images/Q2KDocv2_11b8.pdf

Das, S. (2011). Cleaning of the Ganga. *Journal of the Geological Society of India, 78*(2), 124–130. https://doi.org/10.1007/s12594-011-0073-9

Fan, C., Ko, C. H., & Wang, W. S. (2009). An innovative modeling approach using Qual2K and HEC-RAS integration to assess the impact of tidal effect on River Water quality simulation. *Journal of Environmental Management, 90*(5), 1824–1832. https://doi.org/10.1016/j.jenvman.2008.11.011

Fan, C., Chen, K. H., & Huang, Y. Z. (2021). Model-based carrying capacity investigation and its application to total maximum daily load (TMDL) establishment for river water quality management: A case study in Taiwan. *Journal of Cleaner Production, 291*, 125251. https://doi.org/10.1016/j.jclepro.2020.125251

Gao, L., & Li, D. (2015). A review of hydrological/water-quality models. *Frontiers of Agricultural Science and Engineering, 1*(4), 267–276. https://journal.hep.com.cn/fase/EN/10.15302/J-FASE-2014041

Gopal, B., & Sah, M. (1993). Conservation and management of rivers in India: Case-study of the river Yamuna. *Environmental Conservation, 20*(3), 243–254. https://doi.org/10.1017/S0376892900023031

Gupta, R. C., Gupta, A. K., & Shrivastava, R. K. (2013). Water quality modelling of a stretch of river Kshipra (India). *Nature Environment and Pollution Technology, 12*(3), 511. https://neptjournal.com/upload-images/NL-45-24-24.pdf

Idris, S., Yusuf, A., & Saini, G. (2016). Assessment of surface water quality using QUAL2K software: A case study of River Yamuna, India. *European Journal of Advances in Engineering and Technology, 3*(7), 16–23. https://ejaet.com/PDF/3-7/EJAET-3-7-16-23.pdf

Kamal, N. A., Muhammad, N. S., & Abdullah, J. (2020). Scenario-based pollution discharge simulations and mapping using integrated QUAL2K-GIS. *Environmental Pollution, 259*, 113909. https://doi.org/10.1016/j.envpol.2020.113909

Liu, W. C., & Chen, W. B. (2012). Monitoring sediment oxygen demand for assessment of dissolved oxygen distribution in river. *Environmental monitoring and assessment, 184*, 5589–5599.

Mahalakshmi, R., Sivapragasam, C., & Vanitha, S. (2019). Comparison of BOD5 removal in water hyacinth and duckweed by genetic programming. In *information and communication technology for intelligent systems: Proceedings of ICTIS 2018, volume 1* (pp. 401–408). Springer. https://doi.org/10.1007/978-981-13-1742-2_39

Mangottiri, V., Nambi, I. M., & Suresh Kumar, G. (2011). Application of QUAL2K for assessing waste loading scenario in river Yamuna. *IJAET: International Journal of Advances in Engineering & Technology, 2*(2), 336–344. https://www.researchgate.net/profile/Vasudevan-Mangottiri/publication/291161516_Application_of_Qual2k_for_assessing_waste_loading_scenario_in_river_Yamuna/links/57760e5508ae4645d60d1c8b/Application-of-Qual2k-for-assessing-waste-loading-scenario-in-river-Yamuna.pdf

Martin, J. L., Schottman, R. W., & McCutcheon, S. C. (2018). *Hydrodynamics and transport for water quality modeling*. CRC Press. https://doi.org/10.1201/9780203751510

Mohammad, M. E., Al-Ansari, N., Issa, I. E., & Knutsson, S. (2016). Sediment in Mosul Dam reservoir using the HEC-RAS model. *Lakes & Reservoirs: Research & Management, 21*(3), 235–244. https://doi.org/10.1111/lre.12142

Mummidivarapu, S. K., Rehana, S., & Rao, Y. S. (2023). Mapping and assessment of river water quality under varying hydro-climatic and pollution scenarios by integrating QUAL2K, GEFC, and GIS. *Environmental Research, 239*, 117250. https://doi.org/10.1016/j.envres.2023.117250

Muñoz, D. F., Yin, D., Bakhtyar, R., Moftakhari, H., Xue, Z., Mandli, K., & Ferreira, C. (2022). Inter-model comparison of Delft3D-FM and 2D HEC-RAS for total water level prediction in coastal to inland transition zones. *JAWRA Journal of the American Water Resources Association, 58*(1), 34–49. https://doi.org/10.1111/1752-1688.12952

Nguyen, T. T., Keupers, I., & Willems, P. (2018). Conceptual river water quality model with flexible model structure. *Environmental Modelling & Software, 104*, 102–117. https://doi.org/10.1016/j.envsoft.2018.03.014

Niroobakhsh, M., Musavi-Jahromi, S. H., Manshouri, M., & Sedghi, H. (2012). Prediction of water quality parameter in Jajrood River basin: Application of multi-layer perceptron (MLP) perceptron and radial basis function networks of artificial neural networks (ANNs). *African Journal of Agricultural Research, 7*(29), 4131–4139. https://academicjournals.org/journal/AJAR/article-full-text-pdf/E33990635395.pdf/1000

Palani, S., Liong, S. Y., & Tkalich, P. (2008). An ANN application for water quality forecasting. *Marine Pollution Bulletin, 56*(9), 1586–1597. https://doi.org/10.1016/j.marpolbul.2008.05.021

Pratiwi, D., Sumiarsa, D., Oktavia, D., & Sunardi, S. (2023). Water quality influences self-purification in the Cihawuk and Majalaya segments upstream of the Citarum River, West Java, Indonesia. *Water, 15*(16), 2998. https://doi.org/10.3390/w15162998

Ravinashree, A., Sivapragasam, C., & Vasudevan, M. (2022, June). Developmental strategies for a water quality assessment model with limited datasets–a case study from river Bhavani, India. In *IOP conference series: Earth and environmental science* (Vol. 1032, No. 1, p. 012018). IOP Publishing. https://iopscience.iop.org/article/10.1088/1755-1315/1032/1/012018/meta

Sankararajan, V., Neelakandhan, N., & Chandrasekaran, S. (2017). Modeling of constructed wetland performance in BOD 5 removal for domestic wastewater under changes in relative humidity using genetic programming. *Environmental Monitoring and Assessment, 189*, 1–10. https://doi.org/10.1007/s10661-017-5857-y

Sarda, P. R. (2013). Water quality modeling of river by QUAL2Kw. *International Journal of Advances in Management, Technology & Engineering Sciences, 2*, 13–15. https://www.researchgate.net/publication/282356565_WATER_QUALITY_MODELING_OF_RIVER_BY_QUAL2Kw

Sarkar, A., & Pandey, P. (2015). River water quality modelling using artificial neural network technique. *Aquatic Procedia, 4*, 1070–1077. https://doi.org/10.1016/j.aqpro.2015.02.135

Sharma, K., Mceachern, P., Spafford, M., Zhu, D., & Yu, T. (2009). Spatial variation of sediment oxygen demand in Athabasca River: Influence of water column pollutants. In World Environmental and Water Resources Congress 2009: Great Rivers (pp. 1–12).

Sharma, N. K., Suganya, K., Sivapragasam, C., & Matheswaran, M. (2019, April). Genetic programming modeling for pollutant removal from aerobic bioreactor treating industrial wastewater. In *2019 IEEE international conference on intelligent techniques in control, optimization and signal processing (INCOS)* (pp. 1–4). IEEE. https://doi.org/10.1109/INCOS45849.2019.8951364

Sivapragasam, C., & Muttil, N. (2005). Discharge rating curve extension–a new approach. *Water Resources Management, 19*, 505–520. https://doi.org/10.1007/s11269-005-6811-2

Sivapragasam, C., Natarajan, N., Prabavathy, I., & Ragasudha, D. (2023). Apparent temperature as a modelling substitute for evapotranspiration: Ramanathapuram, Tamil Nadu, India. *Water Practice and Technology, 18*(1), 119–129. https://doi.org/10.2166/wpt.2023.001

Wai, K. P., Chia, M. Y., Koo, C. H., Huang, Y. F., & Chong, W. C. (2022). Applications of deep learning in water quality management: A state-of-the-art review. *Journal of Hydrology*, 128332. https://doi.org/10.1016/j.jhydrol.2022.128332

Ye, H., Guo, S., Li, F., & Li, G. (2013). Water quality evaluation in tidal river reaches of Liaohe River estuary, China using a revised QUAL2K model. *Chinese Geographical Science, 23*, 301–311. https://doi.org/10.1007/s11769-013-0586-9

Yuan, L., Sinshaw, T., & Forshay, K. J. (2020). Review of watershed-scale water quality and nonpoint source pollution models. *Geosciences, 10*(1), 25. https://doi.org/10.3390/geosciences10010025

Zhang, R., Qian, X., Li, H., Yuan, X., & Ye, R. (2012). Selection of optimal river water quality improvement programs using QUAL2K: A case study of Taihu Lake Basin, China. *Science of the Total Environment, 431*, 278–285. https://doi.org/10.1016/j.scitotenv.2012.05.063

Predictive Deep Learning Models for Daily Suspended Sediment Load in the Missouri River, USA

Bibhuti Bhusan Sahoo, Sushinder Kumar Gupta, and Mani Bhushan

Abstract This study aims to evaluate the accuracy of two deep learning models, gated recurrent unit (GRU) and long short-term memory (LSTM), for predicting daily suspended sediment load (SSL) in the Missouri River at Omaha, NE in the United States and compare the results with a traditional time series model, autoregressive integrated moving average (ARIMA) model. The study used daily data of SSL in the Missouri River at Omaha, NE from October 2008 to September 2018. The two deep learning models, GRU and LSTM, were applied to the data and their accuracy was compared with the ARIMA model. The models were trained and tested using a 70–30 split of the data and evaluated based on the coefficient of determination (R^2), root mean square error (RMSE), and mean absolute error (MAE). The results showed that GRU had a better performance with a coefficient of determination of 0.871 and a root mean square error of 0.142, compared to LSTM's coefficient of determination of 0.865 and root mean square error of 0.148. GRU also had a lower mean absolute error of 0.097 compared to LSTM's mean absolute error of 0.101. The study concludes that both GRU and LSTM can be used effectively in SSL modeling. However, GRU requires fewer training constraints and has a faster performance with higher accuracy compared to LSTM, which may need additional data to improve accuracy. The results suggest that GRU can be a promising tool for daily SSL prediction in river basins.

B. B. Sahoo
Department of Agricultural Engineering, Centurion University of Technology and Management, R.Sitapur, Odisha, India

S. K. Gupta (✉)
Department of Groundwater Hydrology Division, National Institute of Hydrology, Roorkee, India

M. Bhushan
Government Engineering College, Khagaria, India

R. Srivastav, P. C. Nayak (eds.), *Applications of Machine Learning in Hydroclimatology*, The Springer Series in Applied Machine Learning,
https://doi.org/10.1007/978-3-031-64403-0_8

1 Introduction

Accurate estimation of suspended sediment load (SSL) in rivers is crucial for effective water resources management. The transportation of sediment in rivers is a complex process, and recent research has focused on developing efficient methods for SSL modeling. Techniques such as Artificial Neural Network (ANN), Support Vector Machines (SVM), and Multiple Layer Perception (MLP) are widely used for this purpose. The use of artificial intelligence (AI) approaches, particularly for modeling complex nonlinear relationships, has grown in recent years due to its effectiveness in situations where the exact relationship between variables is unclear. In addition to AI-based methods, other conventional methods such as regression analysis and hydrodynamic models are also used for SSL estimation. These methods are based on the principle of determining the relationship between SSL and various hydrological and geological parameters such as discharge, rainfall, and soil erosion. However, these conventional methods have limitations, particularly in cases where the relationship between variables is complex or unknown.

It is important to note that the choice of method for SSL estimation depends on the availability of data and the specific requirements of the study. AI-based methods are suitable for situations where large amounts of data are available, while conventional methods are suitable for cases where limited data is available. Furthermore, AI-based methods have the advantage of being able to handle complex nonlinear relationships, while conventional methods may not be suitable in such cases. Therefore, accurate estimation of SSL is crucial for sustainable water resources management, and a combination of both conventional and AI-based methods may be necessary to achieve this goal.

Over the past two decades, the applications of data-driven AI models gained more popularity among the researchers for SSL modeling. Some of the AI-based data-driven models used for SSL modeling include SVM (Aytek & Kişi, 2008; Azamathulla et al., 2013; Alp & Cigizoglu, 2007), ANN (Alp & Cigizoglu, 2007; Cimen, 2008; Gupta et al., 2020; Gupta et al., 2019a), MLP (Gupta et al., 2019b), Adaptive Neuro-Fuzzy Inference System (ANFIS) (Jie & Yu, 2011; Lafdani et al., 2013), Genetic Programming (GP) (Zhu et al., 2007; Melesse et al., 2011), Gene Expression Programming (GEP) (Cigizoglu, 2004), Model Tree (MT) (Khosravi et al., 2018), and many more. The complexity and non-linearity of SSL variables, however, make it difficult for these models to classify the SSL-affecting processes. To improve the accuracy of SSL prediction, more advanced prediction models are therefore still needed.

Deep learning (DL), a cutting-edge technique in the field of artificial intelligence, has proven its capabilities in various areas. Some of the successful applications of the DL (Vafakhah, 2013; Kisi & Zounemat-Kermani, 2016) model include image classification, computer vision, speech recognition, and time series analysis.

Long short-term memory (LSTM) is a popular DL model used for univariate time series modeling. The LSTM is a type of recurrent neural network (RNN) that accumulates extended historical data for collection, representation, and storage in the hidden memory. Also, it updates over time to ensure constancy in the time information in memory for long periods (Kisi et al., 2012). Some of the promising application of LSTM in the field of water resources include streamflow modeling (Li et al., 2019; Zhang et al., 2019), evapotranspiration modeling, water quality modeling, low flow modeling (Barzegar et al., 2020), and rainfall-runoff modeling (Zuo et al., 2020). Another DL model named gated recurrent unit (GRU) was created as a modification of LSTM with a more straightforward structure. Some of the applications of GRU are discussed by (Hu et al., 2020; Sahoo et al., 2019; Li et al., 2020).

To the best of our knowledge, there is no specific evidence for using the GRU method in modeling daily SSL. In this paper, we compare and investigate the prediction performance of GRU and LSTM for SSL modeling. The rest of the article is systematized as follows; Gauging station and data collection information are described in Sect. 2. In Sect. 3, the proposed DL models, i.e., LSTM and GRU, are described. Model development and hyperparameters optimization are discussed in Sect. 4. Model performance evaluation is done in Sect. 5. Section 6 deals with results and discussion where the application for the daily SSL and outcomes are briefed followed by conclusions.

2 Gauging Station and Data Collection

Daily suspended-sediment load concentration data (ton per day) of the Missouri River at gauging station Omaha, Nebraska (Station No: 6610000, drainage area = 836048.162 km^2, Latitude: 41°15′32″N and Longitude: 95°55′20″W) (Fig. 1) functioned by the US Geological Survey (USGS), was used to train and test all the models developed in this study. The daily SSL data from October 2008 to September 2018 (Fig. 2) was acquired from the sediment data portal of USGS (https://cida.usgs.gov/sediment/#).

3 Methodology

4 Long Short-Term Memory (LSTM)

LSTM was presented by Sepp Hochreiter and Jürgen Schmidhuber. The LSTM is a particular type of recurrent neural network (RNN) explicitly designed to evade the long-term dependency problem and to overcome some of the primary shortcomings of RNN, i.e., vanishing gradient and exploding gradient. The structure of LSTM block contains a memory cell and three multiplicative gating units, an input gate

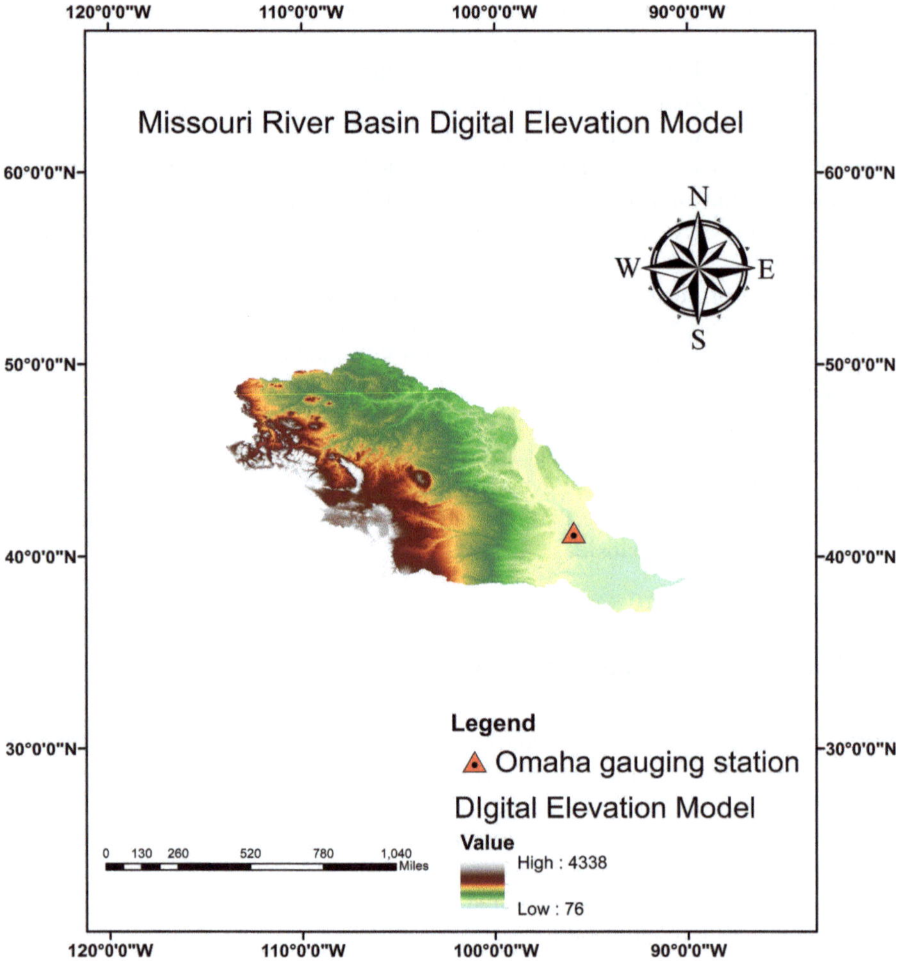

Fig. 1 Missouri River at Omaha gauging station (United States)

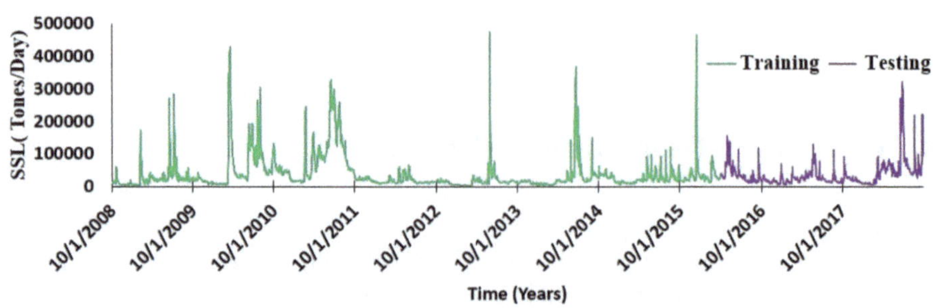

Fig. 2 River SSL time series (October 2018 to September 2018)

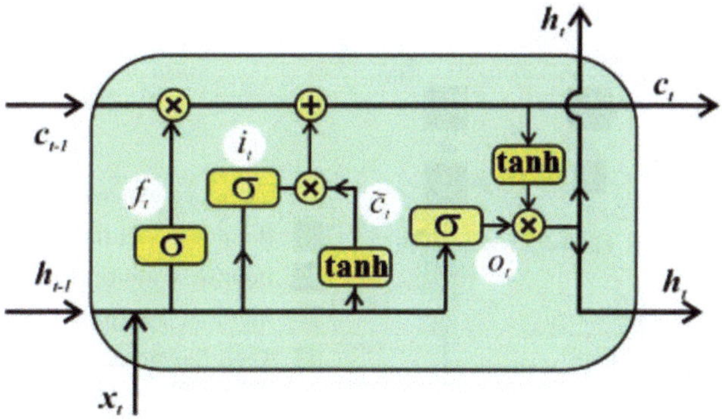

Fig. 3 A simple LSTM architecture

which supplements information to the cell state, a forget gate, which eliminates the information that is no longer essential by the model, and an output gate, which picks the information to be shown as output. Figure 3 shows a simple LSTM architecture.

$$f_t = \sigma\left(W_f \cdot [h_{t-1}, x_t] + b_f\right) \tag{1}$$

$$i_t = \sigma(W_i \cdot [h_{t-1}, x_t] + b_i) \tag{2}$$

$$\tilde{C}_t = \tan h(W_C \cdot [h_{t-1}, x_t] + b_C) \tag{3}$$

$$C_t = f_t C_{t-1} + i_t \tilde{C}_t \tag{4}$$

$$o_t = \sigma(W_o \cdot [h_{t-1}, x_t] + b_o) \tag{5}$$

$$h_t = o_t \tan h(C_t) \tag{6}$$

In these formulas, all *W* variables are weights, and *b* variables are biases, and both are learned during the training phase. More details about LSTM can be found in the papers by and.

5 The Gated Recurrent Unit (GRU)

The GRU is a specific type of LSTM endorsed by Wang et al. (2018). It has shown many advantages over LSTM in many tasks. It trains the model with a lower training timing and fewer constraints than a typical LSTM Yu et al. (2019). Figure 4 shows the structure of a gated recurrent unit network.

The memory state and hidden state of the LSTM are united in GRU as described by Ding et al. (2019). The constraints and control gates in the GRU network are abridged, allowing the GRU to lessen time required to train the model. The GRU framework consists primarily of the reset gate $r_{(t)}$, update gate $z_{(t)}$, fresh memory $\widehat{h}_{(t)}$,

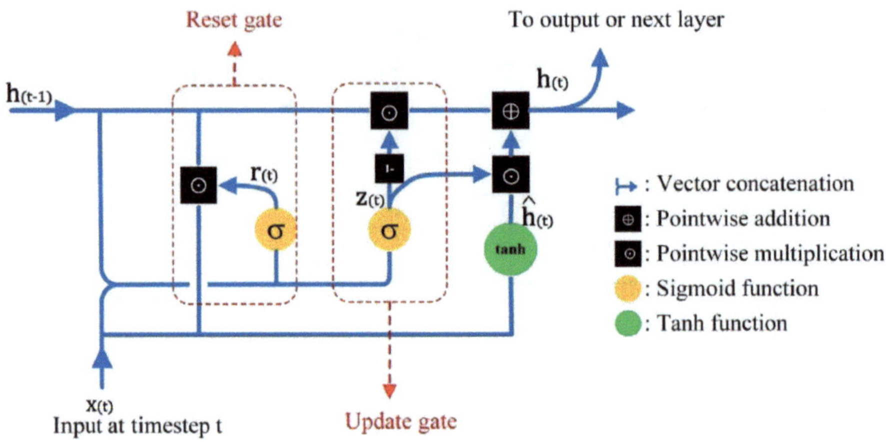

Fig. 4 A simple GRU architecture

and GRU present state $h_{(t)}$. Besides, $h_{(t-1)}$ is the GRU state of the earlier moment and $x_{(t)}$is the contribution of the present condition. Both the reset gate $r_{(t)}$and the update gate $z_{(t)}$ collect the input containing the similar GRU unit's output obtained at an earlier phase $h_{(t-1)}$and input data related to the current time step $x_{(t)}$. Note that W_n and U_n are the weight matrices of input and output data from previous cells, while b_n represents bias vectors with $n \in (r, h, z)$. The above weights and bias together identify the aforementioned gates and fresh memory. Furthermore, $\odot$ and $\oplus$ are pointwise multiplication and pointwise addition, respectively. The most notable improvement in GRU relative to LSTM is considered to be the reduction in memory units and the drop in the number of control gates. Below are the modified gate structures discussed. First, the reset gate $r_{(t)}$is accountable for evaluating how much of the information of the earlier phase $h_{(t-1)}$ can be used in the current state $\widehat{h}_{(t)}$, which is functioned by a "sigmoid" activation function σ within 0 to 1.

$$r_{(t)} = \sigma\left(W_r \cdot x_{(t)} + U_r h_{(t-1)} + b_r\right) \tag{7}$$

Following, the new memory $\widehat{h}_{(t)}$ is updated by brief the fresh input$x_{(t)}$, and the information of reset gate $r_{(t)}$"Hyperbolic tangent" (tanh) is used to produce a vector of fresh memory. It is worth mentioning that the GRU uses hidden cells directly instead of individual memory cells as in LSTM to store memory information

$$\widehat{h}_{(t)} = \tan h\left(U_h\left(r_{(t)} \odot h_{(t-1)}\right) + W_h x_{(t)} + b_h\right) \tag{8}$$

The updated gate $z_{(t)}$, which is the result of GRU simplifying the integration of input and forgotten gates in LSTM into one gate, determines the influence of the previous state $h_{(t-1)}$ on the current state via the sigmoid function. If the value is one, all the previous state information will be transmitted to the current state. In contrast, the current state is determined only by the fresh memory $\widehat{h}_{(t)}$.

$$z_{(t)} = \sigma\left(W_z x_{(t)} + U_z h_{(t-1)} + b_z\right) \tag{9}$$

At last, the current hidden state or output $h_{(t)}$ is determined by the previous state $h_{(t-1)}$ and the fresh memory $\widehat{h}_{(t)}$,

$$h_{(t)} = \left(1 - z_{(t)}\right) \odot h_{(t-1)} \oplus z_{(t)} \odot \widehat{h}_{(t)} \tag{10}$$

6 Autoregressive Integrated Moving Average (ARIMA)

ARIMA is a statistical method that combines three components: autoregression (AR), integration (I), and moving average (MA). Autoregression models the relationship between an observation and a number of lagged observations. Integration is used to handle non-stationary time series, which are time series with a trend or seasonal component. Moving average models the relationship between the errors of previous forecasts and the current observation.

The ARIMA model can be represented by the notation ARIMA(p,d,q), where p is the order of the autoregression component, d is the order of differencing, and q is the order of the moving average component. The order of differencing is used to make the time series stationary, which is a requirement for a well-behaved ARIMA model.

Autoregression (AR) component of ARIMA models the relationship between an observation and a number of lagged observations. The purpose of autoregression is to capture the persistence in the time series data. For example, if the value of a time series at time t is related to its value at time t-1, then the time series is said to have persistence. Moving average models, the relationship between the errors of previous forecasts and the current observation. The purpose of moving average is to capture the randomness in the time series data. For example, if the error of a previous forecast is related to the value of the time series at time t, then the time series is said to have randomness.

The order of moving average, q, refers to the number of lagged errors used in the model. For example, if $q = 1$, then the model will only use the error of the previous forecast to predict the value at time t. If $q = 2$, then the model will use the errors of both the previous and the previous-to-previous forecast to predict the value at time t.

7 Model Development and Hyperparameters Optimization

We trained these models using Keras deep learning library intel core i3-3110m CPU @ 2.40ghz,4GB RAM. While developing a deep learning model, it is crucial to select a suitable loss function and an optimizer. A loss function (or objective function) is one of the essential parameters required to compile a model and a method of evaluating how well the deep learning algorithm models the dataset. There are several different common loss functions to choose from the cross-entropy loss, the mean-squared error (MSE), the Huber loss, and the hinge loss. However, most of the studies use MSE as a loss function related to deep learning models used for regression or prediction. Adam is an adaptive learning rate optimization algorithm that has been designed specifically for training deep learning models and is used as an optimizer in this study. We trained the model with other optimizers like stochastic gradient descent (SGD), Root Mean Square Propagation (RMSprop), and Adaptive Gradient algorithm (AdaGrad), but Adam performed better than others. Optimizing the hyperparameters of any deep learning model is a vital part of any model for better results.

The type of activation mechanism chosen for nodes is another important concern. After a trial-and-error procedure, the Sigmoid and linear activation functions are engaged for hidden and output layers, respectively, for the LSTM model. We tuned the hyperparameters of each model based on their Keras packages. The number of epochs of each model is 500. For LSTM and GRU, we set their units and batch size as 5 and 50, respectively. We used 75% of data for training 25% of the data for testing. From the training set, 33% of the data is randomly used for validation. Further the original SSL data is transformed using the Eq. (11) before training the model, and finally, in the post-processing stage, the predicted value was back-transformed into their actual scale.

$$d' = \ln(d) \tag{11}$$

where $d$is the original SSL value and $d'$is the transformed value.

The analysis was carried out with various time lags as input to develop the models, i.e., four models (M1, M2, M3, M4) for GRU and four models (M1, M2, M3, M4) for LSTM with lag 1, lag 3, lag 6, and lag 12 as shown in Table 1.

where SSL_t denotes the suspended sediment load at time t.

The ARIMA model is developed using auto.arima function from the library forecast in R software.

Table 1 Four models used in the course of the analysis with the corresponding predictor variable as input for GRU and LSTM

M1	$SSL_t = f(SSL_{t-1})$
M 2	$SSL_t = f(SSL_{t-1}, SSL_{t-2}SSL_{t-3})$
M 3	$SSL_t = f(SSL_{t-1}, SSL_{t-2}, SSL_{t-3}, SSL_{t-4}, SSL_{t-5}, SSL_{t-6})$
M 4	$SSL_t = f(SSL_{t-1}, SSL_{t-2}, \ldots\ldots, SSL_{t-12})$

8 Model Performance Evaluation

In the study, the performance of the developed models was evaluated using various statistical matrices and compared their values in the testing set.

Coefficient of determination

$$R^2 = 1 - \frac{\sum_{\mathrm{i}=1}^{n}\left(\mathrm{SSL}_{i(\mathrm{Observed})} - \mathrm{SSL}_{i(\mathrm{Predicted})}\right)^2}{\sum_{\mathrm{i}=1}^{n}\left(\mathrm{SSL}_{i(\mathrm{Observed})} - \overline{\mathrm{SSL}}_{\mathrm{Observed}}\right)^2} \tag{12}$$

Root-mean-square error (RMSE)

$$\mathrm{RMSE} = \sqrt{\frac{1}{n}\sum_{\mathrm{i}=1}^{n}\left(\mathrm{SSL}_{i(\mathrm{Observed})} - \mathrm{SSL}_{i(\mathrm{Predicted})}\right)^2} \tag{13}$$

Mean absolute error (MAE)

$$\mathrm{MAE} = \frac{1}{n}\sum_{\mathrm{i}=1}^{n}\left|\mathrm{SSL}_{i(\mathrm{Observed})} - \mathrm{SSL}_{i(\mathrm{Predicted})}\right| \tag{14}$$

Nash-Sutcliffe Efficiency (NSE)

$$\mathrm{NSE} = 1 - \frac{\sum_{\mathrm{i}=1}^{n}\left(\mathrm{SSL}_{i(\mathrm{Observed})} - \mathrm{SSL}_{i(\mathrm{Predicted})}\right)^2}{\sum_{\mathrm{i}=1}^{n}\left(\mathrm{SSL}_{i(\mathrm{Observed})} - \overline{\mathrm{SSL}}_{\mathrm{Observed}}\right)^2} \tag{15}$$

where $\mathrm{SSL}_{(\mathrm{Observed})}$ is the observed suspended sediment load concentration, $\mathrm{SSL}_{(\mathrm{Predicted})}$ is the predicted suspended sediment load concentration, $\mathrm{SSL}_{(\overline{\mathrm{Observed}})}$ is the mean of the observed suspended sediment concentrations, and n is the number of data points. In regression, the R^2 is a statistical measure of how well the regression predictions approximate the real data points. An R^2 value of 1 indicates that the regression predictions perfectly fit the measured data, while the value of 0 indicates no linear relationship. NSE indicates how well the plot of observed versus simulated data fits the 1:1 line that ranges from minus infinity to 1, with greater values demonstrating superior agreement. The RMSE parameter provides a quantitative indication of the model's absolute error in terms of the units of the variable, with the characteristic that larger errors obtain greater attention than smaller ones. This characteristic can help eliminate techniques with meaningful errors.

9 Results and Discussion

The focus of this research was to explore the efficiency of two DL models, i.e., the GRU and LSTM models to predict SSL. The models were assessed and compared using statistical indicators, including R^2, RMSE, NSE, and MAE, in the training and testing periods. The training and testing results are shown in Table 2 with various lags (lag 1, lag 3, Lg 6, lag 12). The GRU model with input lag 3 performed better among other all models developed with various inputs in the case of GRU. The various statistical scores which support the superiority of the lag 3 GPU model comprise with lowest RMSE (13,747 ton/day), MAE (54,882 ton/day), and higher NSE (0.830), R^2 (0.871). In the case of the LSTM model, the lag 1 model outperformed with RMSE (13,778 ton/day), MAE (5626 ton/day), and higher NSE (0.829), R^2 (0.865). While comparing GPU and LSTM model, GPU performed better with lag 3 with the lowest RMSE error. The scatter plots of the observed and predicted SSL in the testing period (Fig. 5) for GRU with lag 3 and (Fig. 6) for LSTM with lag 1 show that the GRU model performed better than the other models. The training and validation loss of the developed model, i.e., GPU (lag 3) and LSTM (lag 1), is also shown in Figs. 7 and 8 respectively. The developed ARIMA model shows lowest performance RMSE (20,796 ton/day), MAE (8128 ton/day), and higher NSE (0.752), R^2 (0.791) as compared to deep learning models (Table 3).

10 Conclusions

In the present study, we compare and analyze the prediction performance of two deep learning models namely GRU and LSTM for SSL modeling for the Missouri River at Omaha, Nebraska in the United States. To examine the predictive performance of the models, the results were compared to a traditional model ARIMA and various statistical matrices were checked. Evaluation results of the actual performance of the models showed that the GRU model outperformed the LSTM model. Finally, we advocate that the developed deep learning models can be used for modeling the daily suspended sediment load. GRU utilizes less training constraints and thus utilizes fewer memory, performs quicker, and trains faster than LSTM, while LSTM needs additional data to be more accurate. In short, if the available data is more or the precision is very critical, please go to LSTM, while GRU goes for less memory consumption and faster activity.

Table 2 Evaluation of models in SSL prediction on the training and testing period for lag 1, lag 3, lag 6, and lag 12

		Lag 1		Lag 3		Lag 6		Lag 12	
		Training	Testing	Training	Testing	Training	Testing	Training	Testing
GRU	MAE (ton/day)	5817	5928	5760	5488	6487	5844	6645	6038
	RMSE (ton/day)	17,296	13,777	17,370	13,747	18,878	14,392	19,105	14,771
	NSE	0.866	0.840	0.848	0.830	0.808	0.795	0.804	0.780
	R^2	0.896	0.865	0.906	0.871	0.894	0.865	0.891	0.859
LSTM	MAE (ton/day)	5822	5626	6036	5658	6247	5714	6524	5829
	RMSE (ton/day)	17,306	13,778	17,994	14,055	18,313	14,143	18,933	14,184
	NSE	0.865	0.829	0.834	0.819	0.824	0.806	0.829	0.809
	R^2	0.896	0.865	0.9	0.864	0.898	0.867	0.881	0.867

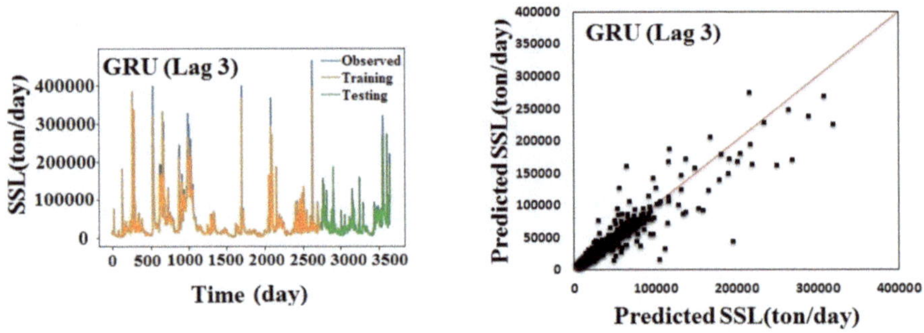

(a) (b)

Fig. 5 (**a**) and (**b**) plots showing observed and predicted SSL during the testing period GRU

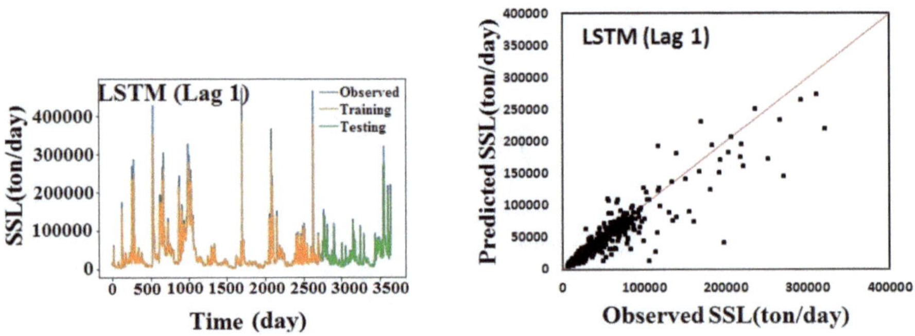

(a) (b)

Fig. 6 (**a**) and (**b**) plots showing observed and predicted SSL during testing period LSTM

Fig. 7 Model loss during training and validation period GRU (lag 3)

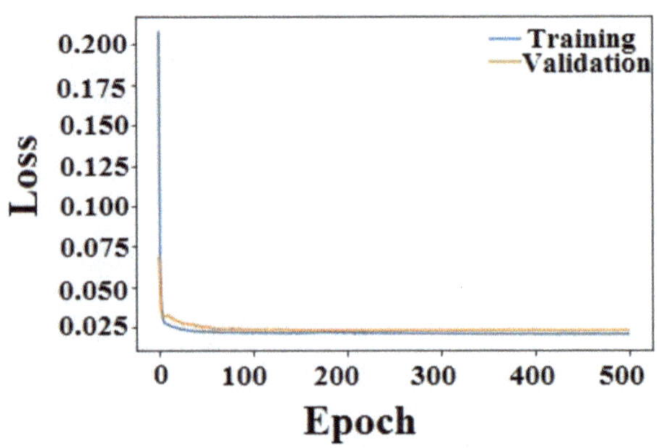

Fig. 8 Model loss during training and validation period LSTM (lag 1)

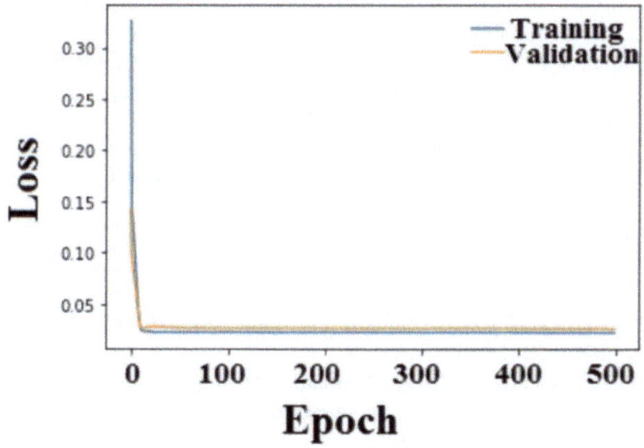

Table 3 Evaluation matrices of ARIMA models in SSL prediction

		Training	Testing
ARIMA	MAE(ton/day)	8817	8128
	RMSE(ton/day)	21,282	20,796
	NSE	0.776	0.752
	R^2	0.763	0.791

Declarations **Acknowledgment**: This manuscript is present in preprint.

Conflicts of interest/Competing Interests: The authors do not have any conflict of interest to disclose.

Availability of Data and Material: All data and models that support the findings of this study are available from the corresponding author upon reasonable request.

Code Availability: All codes that support the findings of this study are available from the corresponding author upon reasonable request.

References

Alp, M., & Cigizoglu, H. K. (2007). Suspended sediment load simulation by two artificial neural network methods using hydrometeorological data. *Environmental Modelling & Software, 22*(1), 2–13.

Aytek, A., & Kişi, Ö. (2008). A genetic programming approach to suspended sediment modelling. *Journal of Hydrology, 351*(3–4), 288–298.

Azamathulla, H. M., et al. (2013). Suspended sediment load prediction of river systems: GEP approach. *Arabian Journal of Geosciences, 6*(9), 3469–3480.

Barzegar, R., Aalami, M. T., & Adamowski, J. (2020). Short-term water quality variable prediction using a hybrid CNN–LSTM deep learning model. *Stochastic Environmental Research and Risk Assessment*, 1–19.

Cigizoglu, H. K. (2004). Estimation and forecasting of daily suspended sediment data by multi-layer perceptrons. *Advances in Water Resources, 27*(2), 185–195.

Cimen, M. (2008). Estimation of daily suspended sediments using support vector machines. *Hydrological Sciences Journal, 53*(3), 656–666.

Ding, M., et al. (2019). A gated recurrent unit neural networks based wind speed error correction model for short-term wind power forecasting. *Neurocomputing, 365*, 54–61.

Gupta, S. K., Tyagi, J., Sharma, G., Jethoo, A. S., & Singh, P. K. (2019a). An event-based sediment yield and runoff modeling using the soil moisture balance/budgeting (SMB) method. *Water Resources Management, 33*, 3721–3741.

Gupta, S. K., Tyagi, J., Singh, P. K., Sharma, G., & Jethoo, A. S. (2019b). Soil moisture accounting (SMA) based sediment graph models for small watersheds. *Journal of Hydrology, 574*, 1129–1151.

Gupta, S. K., Singh, P. K., Tyagi, J., Sharma, G., & Jethoo, A. S. (2020). Rainstorm-generated sediment yield model based on soil moisture proxies (SMP). *Hydrological Processes, 34*(16), 3448–3463.

Hu, Y., et al. (2020). Stream-flow forecasting of small rivers based on LSTM. arXiv preprint arXiv:2001.05681.

Jie, L., & Yu, S. (2011). Suspended sediment load estimate using support vector machines in Kaoping River basin. In *International conference on consumer electronics, communications and networks (IEEE).*

Khosravi, K., Mao, L., Kisi, O., Yaseen, Z. M., & Shahid, S. (2018). Quantifying hourly suspended sediment load using data mining models: case study of a glacierized Andean catchment in Chile. *Journal of Hydrology, 567*, 165–179.

Kisi, O., & Zounemat-Kermani, M. (2016). Suspended sediment modeling using neuro-fuzzy embedded fuzzy c-means clustering technique. *Water Resources Management, 30*(11), 3979–3994.

Kisi, O., et al. (2012). Suspended sediment modeling using genetic programming and soft computing techniques. *Journal of Hydrology, 450*, 48–58.

Lafdani, E. K., Nia, A. M., & Ahmadi, A. (2013). Daily suspended sediment load prediction using artificial neural networks and support vector machines. *Journal of Hydrology, 478*, 50–62.

Li, S., et al. (2019). Deep learning for hyperspectral image classification: An overview. *IEEE Transactions on Geoscience and Remote Sensing, 57*(9), 6690–6709.

Li, W., Kiaghadi, A., & Dawson, C. N. (2020). High temporal resolution rainfall runoff modelling using long-short-term-memory (LSTM) networks. arXiv preprint arXiv:2002.02568.

Melesse, A., et al. (2011). Suspended sediment load prediction of river systems: An artificial neural network approach. *Agricultural Water Management, 98*(5), 855–866.

Sahoo, B. B., et al. (2019). Long short-term memory (LSTM) recurrent neural network for low-flow hydrological time series forecasting. *Acta Geophysica, 67*(5), 1471–1481.

Vafakhah, M. (2013). Comparison of cokriging and adaptive neuro-fuzzy inference system models for suspended sediment load forecasting. *Arabian Journal of Geosciences, 6*(8), 3003–3018.

Wang, Y., Liao, W., & Chang, Y. (2018). Gated recurrent unit network-based short-term photovoltaic forecasting. *Energies, 11*(8), 2163.

Yu, Z., et al. (2019). Deep learning for daily peak load forecasting–a novel gated recurrent neural network combining dynamic time warping. *IEEE Access, 7*, 17184–17194.

Zhang, J., et al. (2019). Medical image classification using synergic deep learning. *Medical Image Analysis, 54*, 10–19.

Zhu, Y.-M., Lu, X., & Zhou, Y. (2007). Suspended sediment flux modeling with artificial neural network: An example of the Longchuanjiang River in the Upper Yangtze Catchment, China. *Geomorphology, 84*(1–2), 111–125.

Zuo, G., et al. (2020). Decomposition ensemble model based on variational mode decomposition and long short-term memory for streamflow forecasting. *Journal of Hydrology*, 124776.

The High-Resolution Statistical Downscaling of Seasonal Rainfall Forecasts Models for Comprehensive Evaluation of Hybrid Gamma Distribution for Districts of West Bengal, India

Aminuddin Ali and Yarunnisha Khatun

Abstract Climate models play a pivotal role in understanding the complex dynamics of the Earth's climate system. The C3S has contributed significantly to our knowledge, providing a wealth of rainfall output from diverse global climate models. However, the inherent spatial resolution of these models often limits their direct applicability for regional scale assessments. This study introduces a Hybrid Gamma Probability Distribution (HGPD) Model designed for comprehensive evaluation of C3S rainfall outputs at local level. The proposed hybrid method incorporates advanced statistical downscaling techniques to bridge the gap between the coarse resolutions of C3S models and the finer scales required for regional assessments. Model development involves the calibration and validation of the downscaling technique using observed rainfall data, ensuring robust performance across diverse climatic regions. The SD model is tailored to capture the intricate spatial patterns of rainfall, enhancing the accuracy of regional scale assessments. To showcase the applicability and effectiveness of the developed SD approach, a grid-level evaluation is conducted using rainfall forecasts. The evaluation encompasses various climate zones, providing a comprehensive overview of the downscaled model's performance in different geographical and climatic contexts.

Key metrics such as spatial distribution, intensity, and variability are analyzed to assess the downscaling model's ability to reproduce observed rainfall patterns.

The findings from this study contribute to advancing our understanding of the reliability and limitations of C3S rainfall outputs on a local scale. The developed downscaling approach offers a valuable tool for researchers and policymakers engaged in regional climate impact assessments, facilitating more accurate and informed decision-making in the face of climate variability and change. This research represents a significant step toward improving the utility of C3S data for

A. Ali
Department of Statistics, Ramkrishna Mission Vidyamandira, Howrah, India

Y. Khatun (✉)
Department of History, Bankura University, Bankura, West Bengal, India

R. Srivastav, P. C. Nayak (eds.), *Applications of Machine Learning in Hydroclimatology*, The Springer Series in Applied Machine Learning,
https://doi.org/10.1007/978-3-031-64403-0_9

regional applications, thereby enhancing our ability to address the complex challenges posed by a changing climate.

1 Introduction

The literature review introduces the context of global warming and its impacts on the hydrological cycle, emphasizing the heightened concerns regarding intensified rainfall related extreme events such as droughts and floods. The need for assessing potential future changes in water resources and hydrological extremes at regional and global scales is highlighted, with the outputs of C3S models serving as a crucial tool for this purpose. However, the coarse representation of orography and other elements in C3S outputs necessitates downscaling procedures for accurate climate change impact assessments (Laprise et al., 2008; Vaittinada Ayar et al., 2016; Von Storch et al., 1993).

Two broad categories of downscaling methods, namely dynamical and statistical, are discussed, each with its own strengths and weaknesses. The focus is on statistical downscaling methods due to their computational efficiency, especially at regional spatial scales such as block and village levels. Existing downscaled C3S datasets are mentioned, mostly concentrated on specific regions like North America (Becker et al., 2014).

The need for a new downscaling technique prompts the proposal of a Bias-corrected Hybrid Gamma Probability Distribution Downscaling method (HGPD). Unlike traditional grid-by-grid statistical downscaling methods, HGPD considers the altitude of randomly generated spatial points, classifying them into groups (Ali, 2020, 2022; Ali & Ghosh, 2022). This approach significantly reduces computation time while incorporating altitude information into the downscaling procedure, thus enhancing the method's performance (Burger & Chen, 2005; Charles et al., 1999; Doblas-Reyes et al., 2013; Giorgi & Mearns, 1999; Gutiérrez et al., 2013).

The objectives of the study are outlined, including the proposal of the HGPD method, verification of its performance against existing downscaled datasets, and the production of a new global downscaled C3S rainfall dataset under different scenarios. The study aims to contribute a new theoretical perspective and a practical downscaling procedure for local rainfall analysis.

The subsequent sections of the article describe the datasets, SD methods, and diagnostic metrics used (Sects. 2, 3, 4 and 5-), present the analysis of results (Sects. 6, 7 and 8), and summarize the conclusions drawn from the study (Sect. 9).

2 Data

The study utilized 53 raw C3S rainfall outputs obtained from the iri with coarse spatial resolution and monthly temporal resolution. These raw outputs were contrasted with the rainfall dataset from the Precipitation Estimation from Remotely Sensed Information using Artificial Neural Network: Climate Data Record (PERSIANN—CDR) [https://chrsdata.eng.uci.edu/] and was used as a surrogate for 24 years observations for Statistical Downscaling and forecast validation, featuring a spatial resolution of 1.0°× 1.0° and monthly temporal resolution. The historical period considered in the study spans from 1993 to 2010, while the validation period extends from 2011 to 2016. The selected periods serve as the basis for evaluating the downscaled C3S rainfall outputs, providing insights into the method's effectiveness in capturing observed rainfall patterns.

3 Statistical Downscaling Method

This study employs SD approaches chosen for their widespread use and potential strength in downscaling coarse-scale models to gridded observations, particularly for representing extremes. Among the selected methods, Bias Correction with Hybrid Gamma Probability Distribution is highlighted, having been applied across WB. In HGPD, a quantile mapping bias correction is applied at the fine scale to address biases caused by the linear combination of daily fields in the constructed analog step. The statistical methods are tested using specific time periods. The study aims to evaluate the techniques for representing hydrologic extremes.

4 Downscaling Performance Evaluation Using Statistical Methods

The study evaluates the modeling accuracy of the HGPD Downscaling method for downscaling rainfall results from two C3S model's rainfall outputs. The performance of the HGPD-based downscaling is compared to a control group, which consists of local-downscaled rainfall products. To assess the rainfall downscaling accuracy of the HGPD method, the study employs two key evaluation metrics: Root Mean Square Error (RMSE) and Pearson correlation analysis. These widely accepted methods in downscaling performance evaluation have been utilized in various studies. The use of these evaluation metrics allows for a comprehensive and quantitative assessment of the HGPD method's performance in reproducing rainfall patterns when compared to others. The study aims to provide insights into the accuracy and reliability of the HGPD downscaling method, contributing to the understanding of its effectiveness in representing rainfall changes in the context of climate modeling.

5 Bias Correction

In the bias correction analysis, the study focuses on defining and utilizing monthly rainfall patterns. Monthly rainfall patterns are classified into four types based on the occurrence time of the maximum rainfall amount within a given year: June to September. The comparison involves assessing the rainfall patterns at the village level.

Surprisingly, the analysis reveals no significant differences in monthly rainfall patterns and amounts under historical and C3S scenarios for all sample points. The historical monthly rainfall differences between observed and the model indices are then utilized to project the spatial and temporal patterns of the monthly rainfall differences in the future. This information forms the basis for generating the bias correction matrix, crucial for refining the downscaled rainfall results.

The bias correction procedure involves several steps:

Monthly historical rainfall pattern analysis of observed and HGPD Downscaled C3S rainfall outputs. Generation of a bias correction vector using the difference between observed rainfall index and two C3S rainfall outputs reprocessed by the HGPD at the sample points.

Generation of the bias correction matrix for the validation period by iterating the bias correction vectors for all sample points.

Bias correction application. The study employs the bias correction matrix to refine downscaled rainfall results. This comprehensive bias correction process enhances the accuracy and reliability of the downscaled rainfall data, providing valuable insights for climate modeling and related analyses.

6 Results and Discussions

The modeling accuracy of the downscaled C3S rainfall outputs is evaluated by comparing the differences between the C3S seasonal rainfall forecasts and observed rainfall. The results highlight the overestimation or underestimation of seasonal rainfall by the original C3S rainfall outputs due to their coarse spatial resolution, emphasizing the necessity of a downscaling procedure. Notably, Meteo France performs better than UKMO in downscaling the original C3S rainfall outputs during 1993–2016 at the local scale. The study's findings underscore the effectiveness of Meteo France in mitigating biases and providing more accurate downscaled rainfall estimates compared to UKMO (Table 1).

Table 1 The details of the correlation coefficient (r) for the global models taken in this study

Year	BD	Meteo France	UKMO	Year	BD	Meteo France	UKMO
1993	0.21	0.88	0.80	2005	0.13	0.54	0.43
1993	0.25	0.79	0.73	2005	0.54	0.88	0.85
1993	0.36	0.86	0.60	2005	0.21	0.91	0.64
1993	0.58	0.37	−0.12	2005	0.60	0.56	0.23
1994	0.52	0.68	0.59	2006	0.49	0.78	0.76
1994	1.00	0.58	0.31	2006	0.83	0.52	0.52
1994	0.60	0.80	0.64	2006	0.72	0.45	0.44
1994	0.16	0.53	0.37	2006	0.45	0.63	0.33
1995	−0.06	0.74	0.65	2007	0.31	0.86	0.76
1995	0.36	0.89	0.83	2007	0.59	0.78	0.54
1995	0.42	0.80	0.76	2007	0.70	0.73	0.42
1995	0.39	0.75	0.71	2007	0.68	0.76	0.34
1996	0.47	0.86	0.87	2008	0.85	0.57	0.69
1996	0.26	0.81	0.80	2008	0.54	0.91	0.69
1996	0.65	0.72	0.52	2008	0.40	0.92	0.76
1996	0.10	0.80	0.57	2008	1.00	0.61	0.33
1997	0.22	0.70	0.66	2009	−0.17	0.54	0.39
1997	0.62	0.80	0.56	2009	0.55	0.69	0.45
1997	0.59	0.50	0.51	2009	0.35	0.92	0.55
1997	0.38	0.91	0.81	2009	0.45	0.38	0.20
1998	0.02	0.52	0.39	2010	0.25	0.79	0.76
1998	0.22	0.76	0.75	2010	0.27	0.14	0.37
1998	0.11	0.81	0.62	2010	0.23	0.80	0.50
1998	0.38	0.67	0.50	2010	0.55	0.84	0.76
1999	0.31	0.90	0.92	2011	0.63	0.45	0.42
1999	0.59	0.91	0.69	2011	0.29	0.23	0.37
1999	0.41	0.86	0.50	2011	0.67	0.66	0.51
1999	0.72	0.56	0.43	2011	0.76	0.45	0.21
2000	0.34	0.91	0.88	2012	0.14	0.62	0.57
2000	0.77	0.75	0.41	2012	0.47	0.22	0.66
2000	0.19	0.83	0.63	2012	0.69	0.77	0.71
2000	0.28	0.48	0.28	2012	0.34	0.81	0.64
2001	0.64	0.82	0.76	2013	0.68	0.76	0.72
2001	0.86	0.78	0.72	2013	0.52	0.48	0.22
2001	0.36	0.78	0.45	2013	0.65	0.59	0.38
2001	0.06	0.69	0.36	2013	0.76	0.48	0.41
2002	0.36	0.89	0.85	2014	0.06	0.65	0.50
2002	0.14	0.78	0.61	2014	0.68	0.34	0.65
2002	0.47	0.48	0.59	2014	0.59	0.87	0.56
2002	0.67	0.69	0.27	2014	0.40	0.90	0.57
2003	0.39	0.86	0.66	2015	0.23	0.50	0.43
2003	0.27	0.78	0.51	2015	0.67	0.41	0.41

(continued)

Table 1 (continued)

Year	BD	Meteo France	UKMO	Year	BD	Meteo France	UKMO
2003	0.65	0.67	0.57	2015	0.32	0.92	0.66
2003	0.38	0.86	0.58	2015	0.29	0.81	0.66
2004	0.36	0.91	0.92	2016	0.16	0.78	0.72
2004	0.14	0.74	0.52	2016	0.46	0.32	0.65
2004	0.72	0.61	0.59	2016	0.78	0.37	0.32
2004	−0.16	0.52	0.58	2016	0.13	0.47	0.37

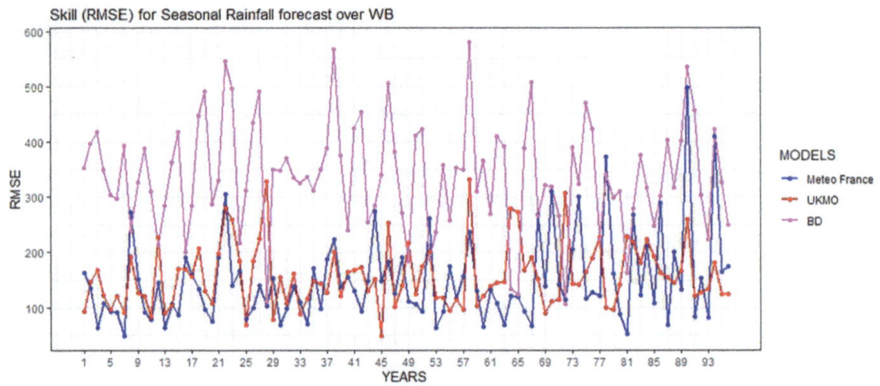

Fig. 1 RMSE for seasonal rainfall forecasts from downscaled Meteo France and UKMO models and before downscaled (BD) models

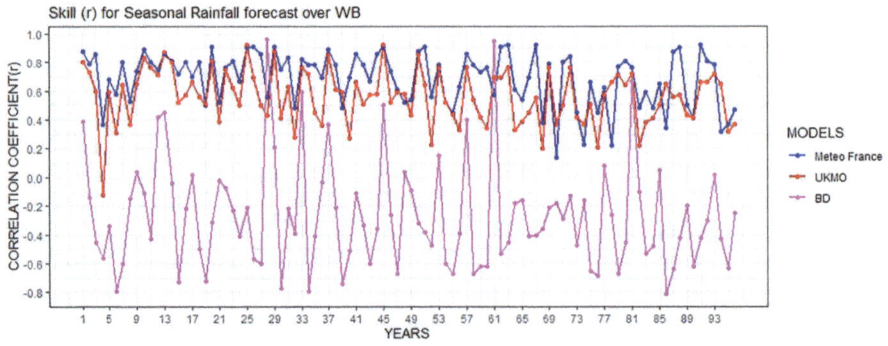

Fig. 2 The correlation coefficient for seasonal rainfall forecasts from downscaled Meteo France and UKMO models and before downscaed (BD) models

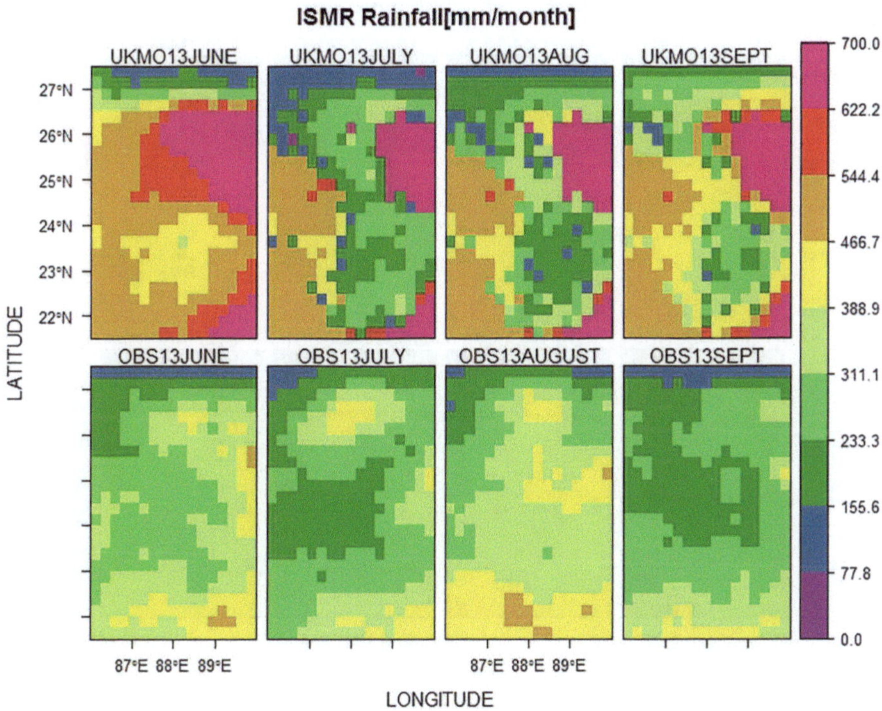

Fig. 3 The high resolution ($0.25^0 \times 0.25^0$) observed (PERSIANN) and forecasted from Statistically Downscaled ISMR forecasts for 2013

7 Intercomparison of RMSE Between Original and Downscaled C3S

The study compares the Root Mean Square Errors (RMSE) between original and downscaled C3S rainfall outputs during the period 1993–2016 at the local scale. The evaluation involves computing RMSE for 2 raw C3S rainfall outputs, in comparison to observed data. The results of the inter-comparison, presented in Table 2 and in Figs. 3, 4, 5 and 6, reveal that HGPD rainfall outputs consistently exhibit the lowest RMSE values across WB, ranging around 49–332 mm for UKMO and 498–49 mm for Meteo France, whereas in contrast, the RMSEs between observed and the original C3S model outputs are notably higher, the RMSE ranges from 58 to 511 mm in case of direct forecasts (BD). These findings demonstrate the superior downscaling performance of UKMO compared to Meteo France.

Furthermore, the study highlights the seasonal raw C3S rainfall outputs, and downscaled C3S rainfall outputs during 1993–2016 at the local scale (Figs. 3, 4, 5 and 6). The results show that downscaled rainfall data closely follow observed for each grid level, indicating a reliable representation of observed rainfall. This further emphasizes the superior and more reliable downscaling methods.

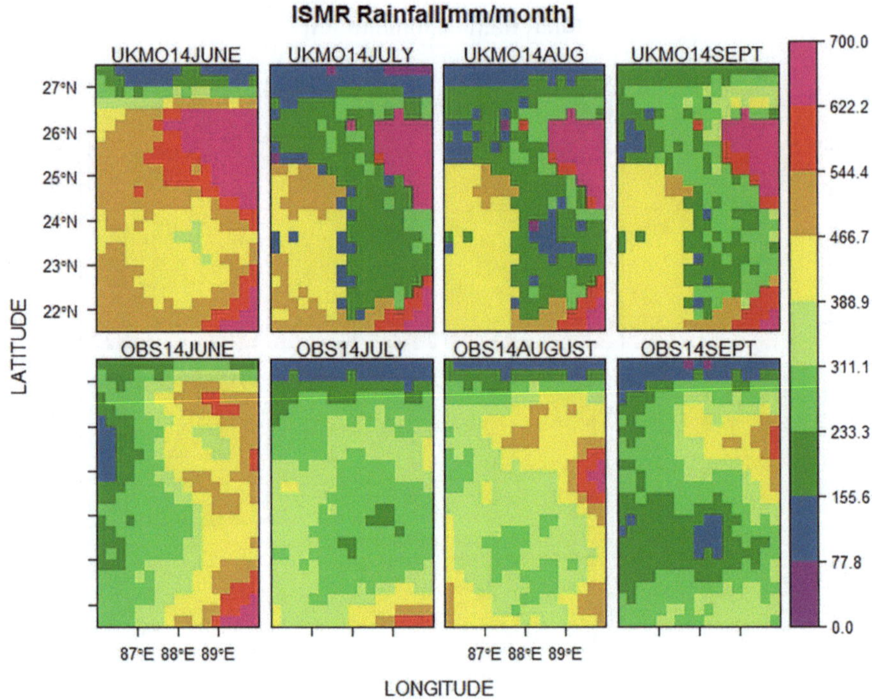

Fig. 4 The high resolution ($0.25^0 \times 0.25^0$) observed (PERSIANN) and forecasted from Statistically Downscaled ISMR forecasts for 2014

8 Spatial Correlation Coefficient (Pearson) Between Observed and Downscaled ECMWF Rainfall

The study assesses Pearson correlation coefficients (r) between downscaled C3S rainfall outputs by HGPD methods and observed data during the period 1993–2016. Table 1 and Fig. 2 illustrate the r values across WB, revealing that the lowest r between HGPD-downscaled and observed rainfall are notably higher (around 0.92 spatial correlation). The significance of r was tested at a 0.05 significance level.

9 Summary and Conclusions

The study introduces the Bias-corrected HGPD Downscaling method and compares its performance between downscaling C3S rainfall outputs and observed. Key conclusions from this research are as follows:

Regional Performance (1993–2016): downscaled C3S rainfall closely matches observed rainfall over WB, while C3S rainfall accurately represents PERSIANN

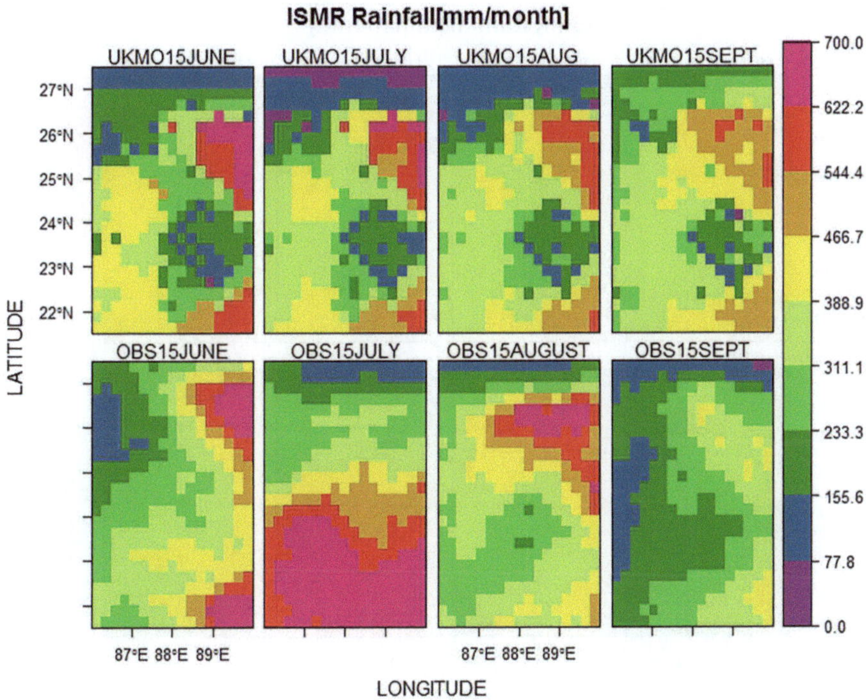

Fig. 5 The high resolution ($0.25^0 \times 0.25^0$) observed (PERSIANN) and forecasted from Statistically Downscaled ISMR forecasts for 2015

changes up to the district level. HGPD successfully does not avoid underestimating in some parts of WB.

Validation Period (2011–2016): During the validation period, HGPD consistently outperforms. The average Root Mean Square Errors (RMSE) between C3S rainfall and PERSIANN are promising (Fig. 1). HGPD shows higher correlation with PERSIANN.

In summary, the study proposes HGPD as an enhanced downscaling technique for C3S and potentially C3S rainfall outputs up to the district level, offering improved accuracy in representing spatial and temporal rainfall patterns compared.

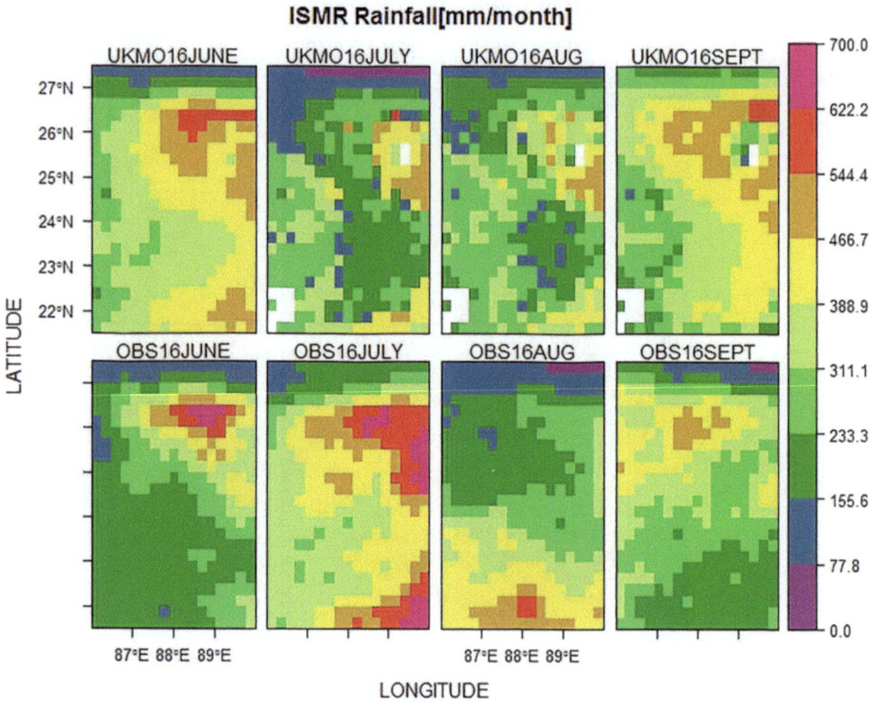

Fig. 6 The high resolution (0.25^0 x 0.25^0) observed (PERSIANN) and forecasted from Statistically Downscaled ISMR forecasts for 2016

Table 2 The details RMSE description of all the global models taken in this study

RMSE	BD	Meteo France	UKMO	RMSE	BD	Meteo France	UKMO
1993	306	163	93	2005	139	111	216
1993	269	136	148	2005	250	106	124
1993	204	64	168	2005	196	93	175
1993	124	107	123	2005	139	261	201
1994	202	94	97	2006	184	63	118
1994	125	92	121	2006	151	93	118
1994	149	49	91	2006	62	174	95
1994	159	273	192	2006	139	116	114
1995	290	152	128	2007	295	157	96
1995	238	91	120	2007	376	236	332
1995	154	78	83	2007	77	130	103
1995	286	145	226	2007	168	65	121
1996	186	64	89	2008	247	133	139
1996	234	106	104	2008	249	107	145
1996	174	87	169	2008	216	69	147
1996	124	191	170	2008	58	121	280
1997	285	160	156	2009	112	120	272
1997	268	134	208	2009	206	92	168
1997	152	96	131	2009	223	67	191
1997	154	76	107	2009	147	267	151
1998	272	191	197	2010	270	139	90
1998	448	305	281	2010	323	310	111
1998	291	140	260	2010	103	141	114
1998	100	166	184	2010	74	114	308
1999	223	79	68	2011	325	206	143
1999	243	99	184	2011	313	300	143
1999	285	140	225	2011	219	116	165
1999	147	102	328	2011	81	127	189
2000	295	154	79	2012	193	121	227
2000	159	69	155	2012	385	372	99
2000	157	98	107	2012	67	162	96
2000	193	139	162	2012	106	88	141
2001	238	109	88	2013	148	53	228
2001	118	69	116	2013	282	268	218
2001	105	172	149	2013	97	123	181
2001	146	98	143	2013	79	211	224
2002	328	188	127	2014	233	108	192
2002	373	223	201	2014	302	289	163
2002	99	138	121	2014	129	68	156
2002	75	156	165	2014	130	201	145
2003	272	131	167	2015	227	133	167
2003	229	93	173	2015	511	498	259

(continued)

Table 2 (continued)

RMSE	BD	Meteo France	UKMO	RMSE	BD	Meteo France	UKMO
2003	57	148	130	2015	151	84	121
2003	169	274	152	2015	89	154	128
2004	291	148	49	2016	162	82	134
2004	329	182	253	2016	423	409	181
2004	70	126	101	2016	78	164	123
2004	180	190	141	2016	120	175	124

References

Ali, A. (2020). Statistical downscaling for improving global model rainfall forecasts of seasonal rainfall over north-east India. *International Journal of Research and Analytical Reviews, 7*(2), 220–236.

Ali, A. (2022). The high resolutions improved global model rainfall forecasts of seasonal rainfall using statistical downscaling methods over South Asia. *Aligarh Journal of Statistics, 42*, 27–40.

Ali, A., & Ghosh, T. (2022). Statistical downscaling method for improving global model rainfall forecasts of seasonal rainfall over West Bengal (WB), India. In *Climate, Environment and Disaster in Developing Countries* (pp. 61–87). Springer.

Becker, E., den Dool, H. V., & Zhang, Q. (2014). Predictability and forecast skill in NMME. *Journal of Climate, 27*(15), 5891–5906.

Burger, G., & Chen, Y. (2005). Regression-based downscaling of spatial variability for hydrologic applications. *Journal of hydrology, 311*(1-4), 299–317.

Charles, S. P., Bates, B. C., & Hughes, J. P. (1999). A spatiotemporal model for downscaling precipitation occurrence and amounts. *Journal of Geophysical Research: Atmospheres, 104*(D24), 31657–31669.

Doblas-Reyes, F. J., García-Serrano, J., Lienert, F., Biescas, A. P., & Rodrigues, L. R. L. (2013). Seasonal climate predictability and forecasting: Status and prospects. *Wiley Interdisciplinary Reviews: Climate Change, 4*(4), 245–268.

Giorgi, F., & Mearns, L. O. (1999). Introduction to special section: Regional climate modeling revisited.

Gutiérrez, J. M., San-Martín, D., Brands, S., Manzanas, R., & Herrera, S. (2013). Reassessing statistical downscaling techniques for their robust application under climate change conditions. *Journal of Climate, 26*(1), 171–188.

Laprise, R., De Elia, R., Caya, D., Biner, S., Lucas-Picher, P., Diaconescu, E., Leduc, M., Alexandru, A., Separovic, L., & for Regional Climate Modelling, C. N., & Diagnostics. (2008). Challenging some tenets of regional climate modelling. *Meteorology and Atmospheric Physics, 100*, 3–22.

Vaittinada Ayar, P., Vrac, M., Bastin, S., Carreau, J., & D´equ´e, M., & Gallardo, C. (2016). Intercomparison of statistical and dynamical downscaling models under the euro-and med-cordex initiative framework: Present climate evaluations. *Climate Dynamics, 46*, 1301–1329.

Von Storch, H., Zorita, E., & Cubasch, U. (1993). Downscaling of global climate change estimates to regional scales: An application to Iberian rainfall in wintertime. *Journal of Climate, 6*(6), 1161–1171.

Prediction of Rainfall in One of the Wettest Regions in India Using Machine Learning Methods

Vishal Singh, Japjeet Singh, Sanjay Kumar Jain, and Pushpendra Kumar Singh

Abstract Machine learning techniques are now extensively being used for prediction of long-term and short-term rainfall time series datasets because the lack of data and availability of recorded daily rainfalls may affect the feasibility of a larger range of rainfall-based studies in light of repercussions from climate change and extreme hydro-meteorological phenomena. A daily gridded rainfall time series dataset (2007–2020) over Mizoram state is constructed using data machine learning methods such as Support Vector Regression and Random Forest. For this purpose, the open sources and observed gridded rainfall datasets CHIRPS, APHRODITE, IMDAA re-analysis, and IMD have been utilized. In this study, the statistical evaluation methods have been employed and tested over selected rainfall grids to enhance the accuracy of the datasets. Quantile-quantile (Q-Q) plots were also used to test the accuracy of each dataset in the case of extreme rainfalls with respect to IMDAA gridded rainfall (i.e., IMD Re-analysis), which has been taken as the observed/reference dataset. In this study, the quantile mapping and linear scaling bias correction methods are utilized to correct the rainfall datasets. The predicted rainfall datasets have been compared produced by SVR and RF, and the RF-based predicted rainfall datasets have performed superior than SVR datasets in terms of bias and extremity.

V. Singh (✉)
Centre for Cryosphere and Climate Change Studies, National Institute of Hydrology, Roorkee, India
e-mail: vishal18.nihr@gov.in

J. Singh
Department of Civil Engineering, Indian Institute of Technology Roorkee, Roorkee, India

S. K. Jain
Centre for Cryosphere and Climate Change Studies, National Institute of Hydrology, Roorkee, India

Department of Civil Engineering, Indian Institute of Technology Roorkee, Roorkee, India

P. K. Singh
Water Resources Systems Division, National Institute of Hydrology, Roorkee, India

R. Srivastav, P. C. Nayak (eds.), *Applications of Machine Learning in Hydroclimatology*, The Springer Series in Applied Machine Learning,
https://doi.org/10.1007/978-3-031-64403-0_10

1 Introduction

In a data scarce region like northeast India, the density of gauges and the long-term availability of rainfall time series are rather limited. The availability of universally acknowledged high-resolution gridded rainfall datasets CHIRPS (Climate Hazards Group Infrared Precipitation with Station Data) (Funk et al., 2015), TRMM (Tropical Rainfall Measuring Mission) (Simpson et al., 1996), GPM (Global Precipitation Measurement), PERSIANN (Precipitation Estimation from Remotely Sensed Information using Artificial Neural Network), and APHRODITE (Asian Precipitation—Highly-Resolved Observational Data Integration Towards Evaluation) (Yatagai et al., 2012), etc., provides a viable source to assess the rainfall variability and pattern in different parts of the world. The resolution and time step of spatial precipitation data products impact the accuracy and outcome of any precipitation-based analysis (Wagner et al., 2012). The reliability of these gridded rainfall datasets has been explored around the world which provides valuable information for long-term assessment of rainfall variability even at a regional scale (Singh et al., 2019). Still, their temporal and spatial availability restricts assessment of long-term impact of rainfall at higher resolution spatial scale. Floods and similar extreme events are consequences of climate change that are expected to occur more frequently and have catastrophic effects in years to come.

Rainfall forecasting has been around for years using traditional methods that employ statistical techniques to assess the correlation between rainfall, geographic coordinates (such as latitude and longitude), and other atmospheric factors (like pressure, temperature, wind speed, and humidity). Insights from previous work/studies recommend that machine learning algorithms perform well regarding rainfall/precipitation prediction accuracy and timeliness. Therefore, it is imperative to examine various classification algorithms to establish the best-performing techniques for predicting rainfall in the wettest regions of India.

2 Machine Learning Methods

Rainfall forecasting has gained utmost research relevance in recent times due to its complexities and persistent applications such as flood forecasting and monitoring of pollutant concentration levels, among others. Existing models use complex statistical models that are often too costly, both computationally and budgetary, or are not applied to downstream applications. Therefore, approaches that use Machine Learning algorithms in conjunction with time-series data are being explored as an alternative to overcome these drawbacks. To this end, this study presents a comparative analysis using simplified rainfall estimation models based on conventional Machine Learning algorithms and Deep Learning architectures that are efficient for these downstream applications. Models based on LSTM, Stacked-LSTM, Bidirectional-LSTM Networks, XGBoost, and an ensemble of Gradient Boosting Regressor,

Linear Support Vector Regression, and an Extra-trees Regressor were compared in the task of forecasting hourly rainfall volumes using time-series data. Machine learning methods can be used to predict rainfall by analyzing historical weather data and other relevant variables. A few common machine learning techniques that can be applied in rainfall prediction:

Regression models can be used to predict the amount of rainfall based on various hydro-meteorological variables such as temperature, humidity, wind speed, and other geographical factors. Algorithms like linear regression, decision trees, or random forests can be employed for the analysis of rainfall prediction (Zainudin et al., 2016) (Misra et al., 2020). Time series analysis of the rainfall data often exhibits temporal patterns, and time series analysis techniques can capture these patterns. Several mathematical models like autoregressive integrated moving average (ARIMA) or seasonal decomposition of time series (STL) can be well utilized to forecast future rainfall based on the available long-term time series historical information (Dwivedi et al., 2019) (He et al., 2022). However, it is important to note that the accuracy of rainfall predictions depends on the availability and quality of data, the selection of relevant features, and the appropriate choice and tuning of the machine learning algorithm. Additionally, domain expertise and understanding of meteorological variables can help improve the accuracy of rainfall predictions.

2.1 LSTM

Long short-term memory (LSTM) is a recurrent neural network (RNN) architecture that has proven to be effective in modeling sequential data, such as time series analysis of rainfall (Chen et al., 2022). LSTM networks are particularly well-suited for rainfall prediction because they can capture long-term dependencies and temporal patterns in the data. In the context of rainfall prediction, LSTM can be trained to forecast future rainfall based on historical rainfall data and other relevant meteorological variables. In rainfall prediction, LSTM can be trained to forecast future rainfall based on the historical rainfall data and other relevant hydro-meteorological variables. The following steps can be utilized to predict rainfall using LSTM:

(i) Data preparation: The historical rainfall data, along with other hydro-meteorological variables such as temperature, humidity, wind speed, etc., can be organized into sequences or time steps. Each time step contains multiple features. For example, a sequence may include rainfall measurements and meteorological data for the past few days.
(ii) Network architecture: The LSTM network is designed to learn the temporal patterns in the input data. It consists of LSTM cells that process the sequential input and maintain a hidden state, which retains information from previous time steps. The output of the LSTM cells can be fed into fully connected layers for final rainfall predictions.

(iii) Training: The LSTM model is trained on the prepared dataset using an optimization algorithm such as gradient descent. During training, the model learns to minimize the difference between predicted rainfall and actual rainfall by adjusting the weights of the network. The training process involves forward propagation, backpropagation, and weight updates.
(iv) Forecasting: Once the LSTM model is trained, it can be used to predict future rainfall. The rainfall predictions can be performed step-by-step, where each prediction is used as input for the next time step.

2.2 *Support Vector Regression*

Support vector regression is a supervised machine learning algorithm which is also helpful in classification and regression problems (Vapnik, 1999) (Mariette & Rahul, 2015). The support vector algorithm's main aim is to identify a hyperplane in an N-dimensional space that classifies the data. The dimension of the hyperplane is determined by the number of features. As the number of features increases the dimension also increases and the classifying plane or line becomes more complex. The samples of different feature which determines the hyperplane position are known as support vectors.

The best hyperplane is the one which has highest margin with respect to feature classes. The parameters to be tuned in support vector regressor are kernel, regularization parameter (C) which suggests the threshold for the misclassification of the samples, higher value allows less sample to misclassify while lower value allows more number of samples to misclassify (Fig. 1).

SVR has many kernels for various applications, including Linear, Polynomial, Radial Basis Function (RBF), and Sigmoid. Linear kernel is 1-D and the most fundamental kernel type in SVR. It is utilized when the data is linearly separable, or can be split by a single line. It is often employed when a dataset has a significant number of characteristics. A kernel function widely used with SVRs and other kernelized models, the polynomial kernel, assesses the similarity of vectors in a feature space over polynomials of the actual variables and facilitates the learning of non-linear models. The polynomial kernel considers combinations of these properties in addition to the individual features of the input samples to assess their degree of similarity. Within the context of regression analysis, these pairings are known as interaction features. The (implicit) feature space of a polynomial kernel is identical to that of polynomial regression; however, there is no combinatorial explosion in the number of parameters that must be learned. When the input features contain binary values, the features correspond to logical coexistence of input features. RBF is a non-parametric model, hence its complexity increases as training set size increases. A RBF kernel SVM is more expensive not only in terms of training expenses, but also in terms of sustaining the kernel matrix and estimating the data into an "infinite" higher dimensional space where it is linearly distinguishable.

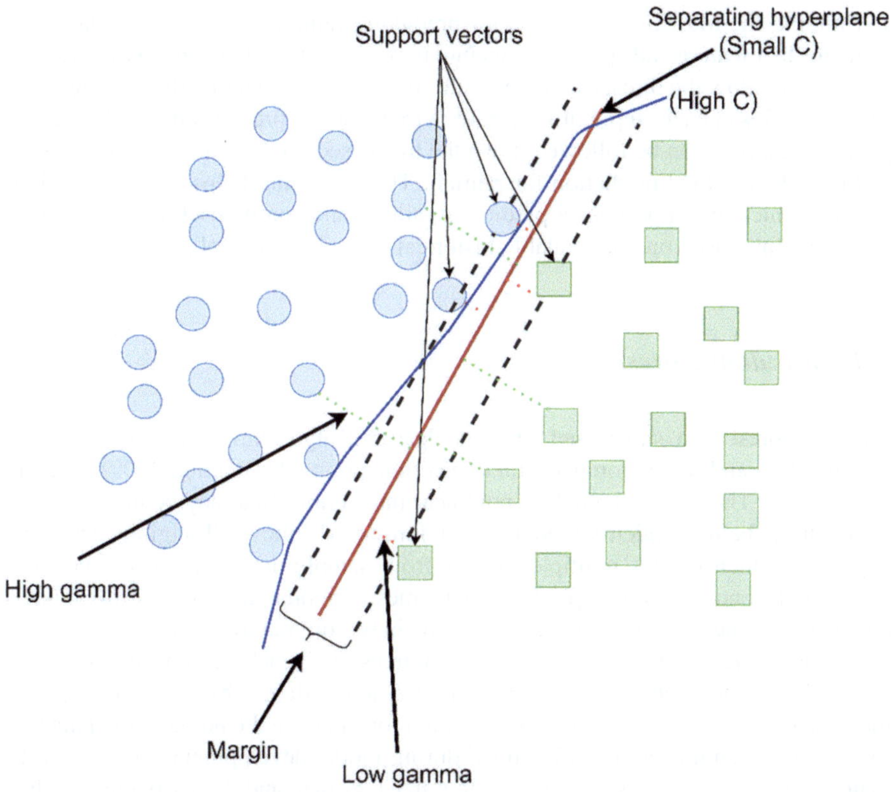

Fig. 1 Support vector machine algorithm

Next parameter is gamma which suggests the influence of a sample based on its position, high gamma value considers only points near to the hyperplane while low gamma value considers far points from hyperplane. Lastly, epsilon parameter indicates penalty assigned during training the model to predicted points from an epsilon distance from the original values which means the larger the epsilon the more error is added to the predicted value. The kernel used in this study is radial basis function kernel (rbf), regularization parameter is kept at 10000 and epsilon is at 0.01 which has given the best solution.

2.3 Decision Tree

The Decision Tree (DT) algorithm is a method used to predict rainfall and many studies showed a well results (Wei et al., 2020). DT works by creating a tree-like structure that makes decisions based on different features. The algorithm starts by looking at historical meteorological variables such as rainfall, temperature, wind, etc.

and finding the most important one for predicting rainfall. It then sets a threshold value for that feature and splits the data into two groups based on whether the feature is above or below the threshold. It continues this process, selecting different features and thresholds at each step, until it creates a tree that can predict rainfall. To make a prediction, a path can be followed down the tree based on the values of the features, and it leads to a final prediction for rainfall. This algorithm helps in understanding features which are important in predicting rainfall and can be useful in forecasting future rainfall based on the available historical data (Tharun et al., 2018).

2.4 Random Forest

Random forest is a supervised machine learning algorithm which is helpful in classification and regression decision tree combinations (Grömping, 2009) (Svetnik et al., 2003). The approach enables nonlinear interactions that may be inferred from data without being explicitly modeled. Random forest follows "Bagging" ensemble approach, which creates multiple models from sample data. Then each model is trained independently which produces outcomes. The outcomes of multiple models are averaged and final result have been made based on majority vote.

Random forest combines several decision trees, with each tree constructed using randomly selected subsets of training samples and variables by means of replacement, as many trees are combined making it a forest hence the name as explained in Fig. 2. Some parameters must be tuned during model development such as n_estimator which is a number of regression trees grown and hyper-parameter like max_depth which is the maximum depth to which trees can grow. Random forest was discovered to have strong predictive performance, low overfitting, and simple interpretability.

3 Study Area and Data Utilization

The present study was performed over the Northeastern region which corresponded to the Latitudes 22.0° to 26.0° and Longitudes 92.0° to 94.5°, which covers Mizoram and several parts of Assam, Tripura, Manipur, Meghalaya, and Nagaland (Fig. 3).

The selected study area comprises the parts of three river basins viz. Barak, Minor rivers draining into Bangladesh, and Minor rivers draining into Myanmar. In this study area, the hills are extremely rugged and steep; the ranges are running in North-South direction, leaving some plains scattered occasionally here and there. The datasets which are used in the study have shown in Table 1.

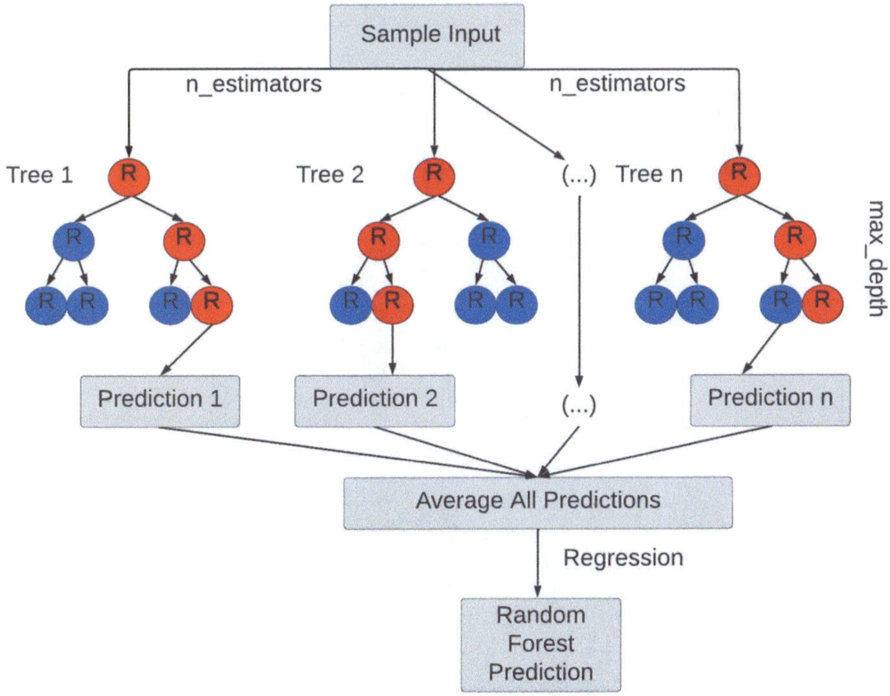

Fig. 2 Random forest algorithm schematic diagram

4 Results and Discussion

4.1 Evaluation of Best Dataset

Figure 3 shows the seasonal variability in mean rainfalls (1985–2015) at 25 km^2 resolution scale among all the selected datasets. Meghalaya receives the largest amount of rainfall during monsoon season and this region can be categorized as the highest rainfall area around the world. The other datasets show underestimation of rainfall in many regions w.r.t. IMDAA. As per mean rainfall distribution, the CHIRPS and APHRODITE rainfall datasets performed well w.r.t. IMDAA and therefore selected to predict the high-resolution (5km^2) rainfall dataset (Fig. 4).

For statistical evaluation of these datasets, the boxplots have been generated by selecting 13 random grids and the statistical evaluation functions viz. R^2, RMSE, MAE, MSE, AIC, and HQC are employed as shown in Fig. 5.

HQC and AIC plot show relatively lower values as compared to IMD and PRINCETON. MSE, RMSE and MAE are also on lower side for APHRODITE and CHIRPS. The R^2 values are found to be higher for APHRODITE (0.63) and

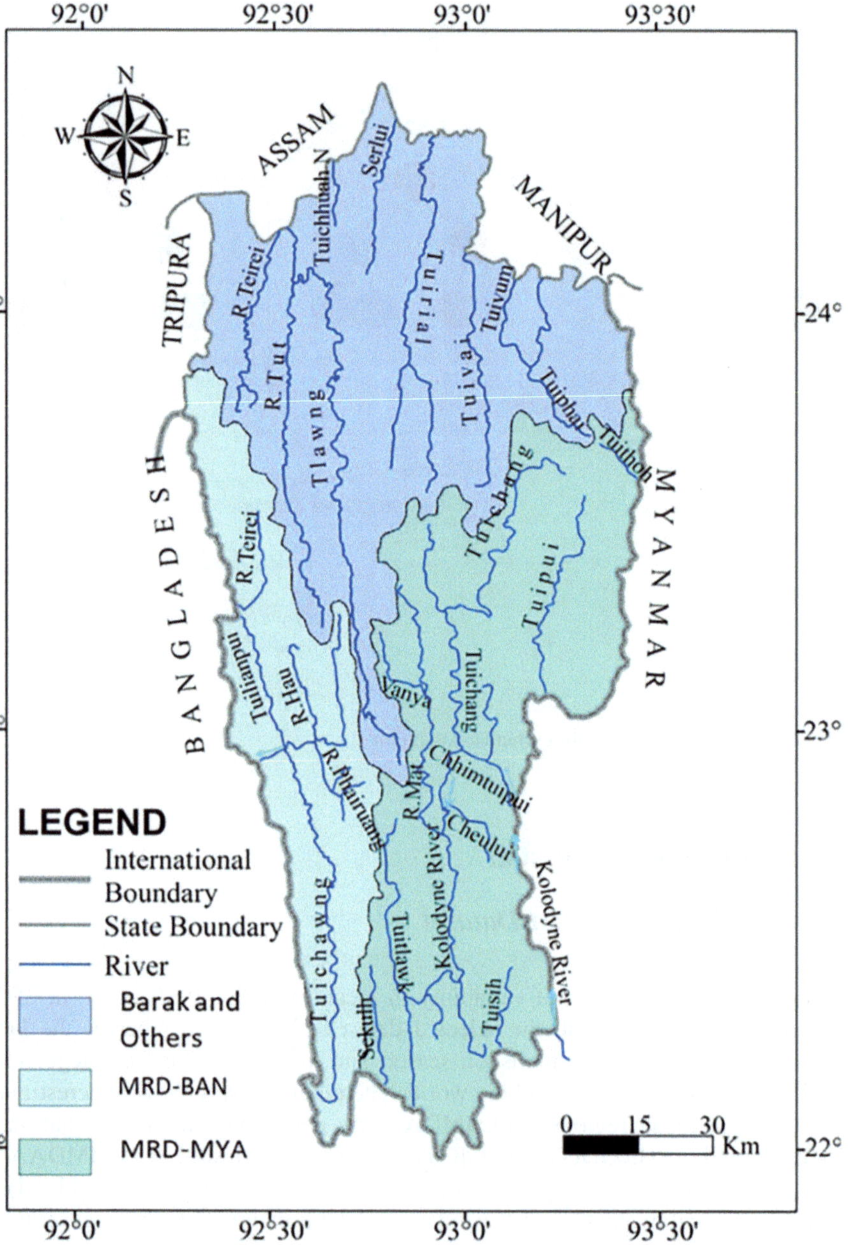

Fig. 3 Map of the selected study region

Table 1 Datasets used in the study

SI. No.	Data-sets/name	Availability/source/resolution	Time-period
1	Digital Elevation Model	SRTM & CARTOSAT Merged – 30 m	-
2	Daily Rainfall Gridded (Observed)	IMD/IMDAA – at 25 km^2/12 km^2	1901-2020/ 1979-2020
3	Daily Temperature (Observed)	IMD – at 100 km^2 & IMDAA – at 12 km^2	1951-2016 & 1979–2020
4	Satellite-based open sources Datasets	CHIRPS, APHRODITE, PRINCETON, at 5 km^2 & 25 km^2	1991-2018

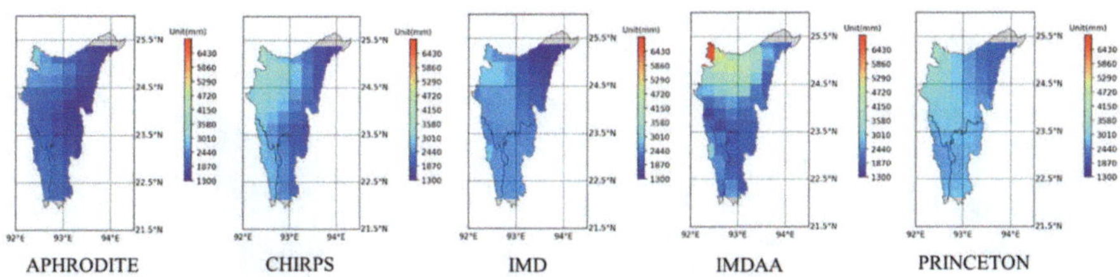

Fig. 4 Averaged annual long term scenario of different selected rainfall datasets

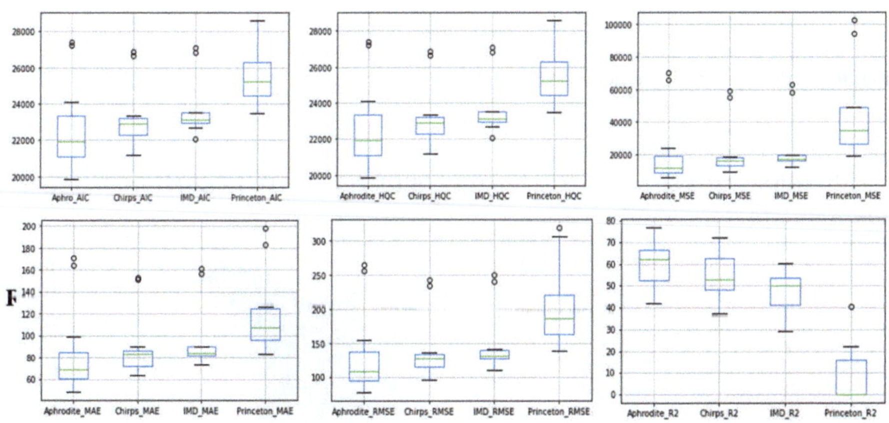

Fig. 5 Boxplots for selected grids

CHIRPS (0.54) as compared to IMD (0.50) and PRINCETON. PRINCETON has shown worst results among all datasets for each statistical parameter. RMSE, MSE, and MAE were quite higher for IMD. Based on the analysis, it was seen that APHODITE and CHIRPS performed better and were considered as the best datasets for further analysis.

4.2 Prediction of Rainfall Datasets

For the prediction of rainfall using RF and SVR methods, the IMDAA, CHIRPS, and APHRODITE rainfall datasets (1989–2015) were utilized (Figs. 6 and 7). In RF and SVM, the time series datasets were divided into training (1989–2007) and testing (2007–2015) datasets. Figure 6 shows the results of predicted rainfall (2007–2015) by RF and SVR. For the assessment of predicted rainfalls, the average annual mean plots (2007–2015) of predicted rainfall datasets and IMDAA datasets are compared as shown in Figs. 7 and 8.

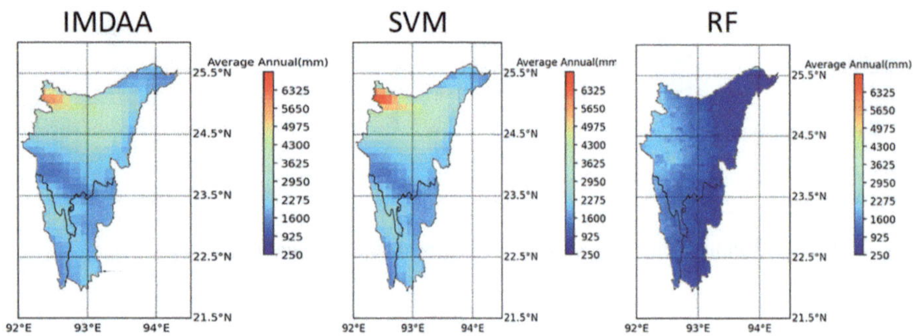

Fig. 6 Predicted rainfall datasets (2007–2015)

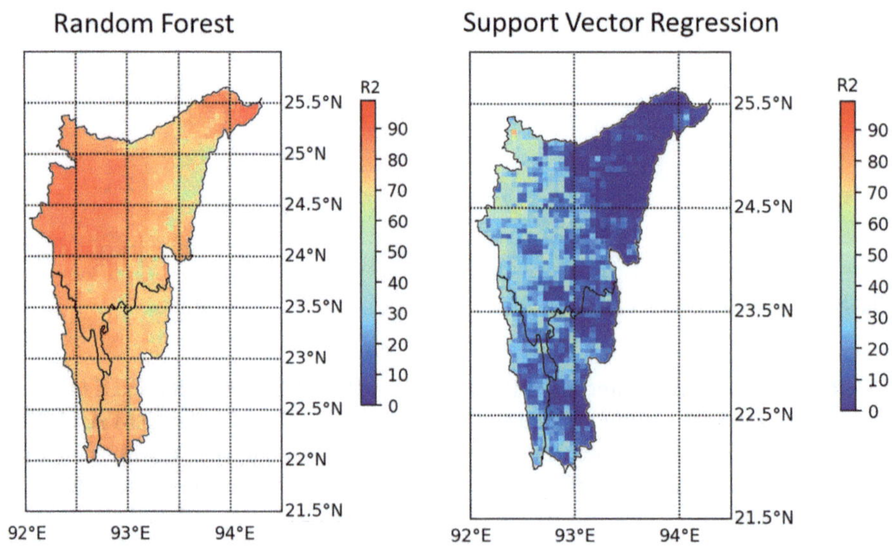

Fig. 7 Testing of RF and SVR predicted rainfall w.r.t. IMDAA data

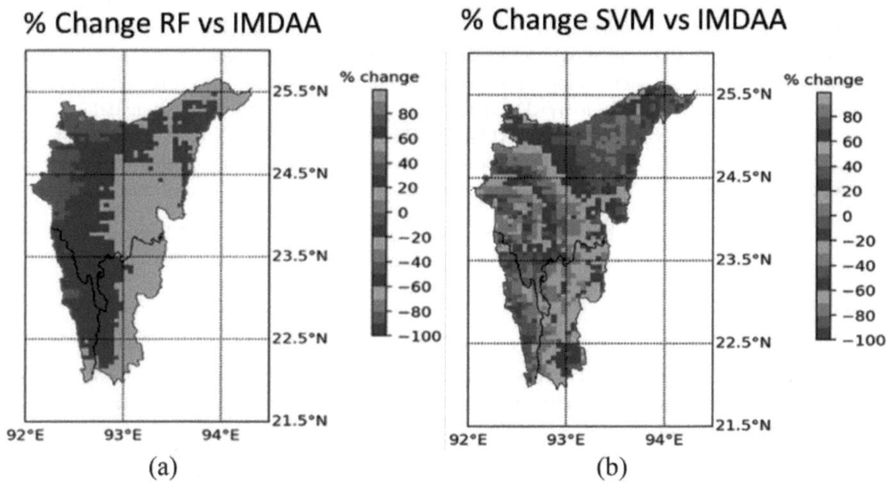

Fig. 8 (**a**) Evaluation of RF and SVR-based predicted rainfall w.r.t. IMDAA rainfall dataset during (2001–2007) and (**b**) shows percentage of change w.r.t. IMDAA

As per the comparative assessment, the RF-based rainfall distribution and pattern look closer to the IMDAA and Assimilated RF. Based on the annual rainfall plots (Figs. 6 and 7), the SVR was not performed well and it is not able to capture the extreme high rainfall values (>4500 mm). Here, RF is found able to capture extreme rainfall values. However, as per the comparative assessment of SVR vs IMDAA and RF vs IMDAA, the RF-based annual mean plot looks more prominent w.r.t. IMDAA than SVR.

5 Conclusion

In this study, the RF and SVR machine learning methods have been utilized to predict the rainfall datasets, which was found very useful for the comparison and evaluation of assimilated rainfall product and other sources rainfall datasets. The overall performance of these two methods viz. RF and SVR in predicting rainfall on the continuous time series is found satisfactory. The strength of the applied SVR approach is its capability to capture the intrinsic non-linear characteristics between different rainfall datasets. Based on the assessment of predicted rainfall datasets (as per RF and SVR), the RF algorithm is performed well. It is clearly observed that in cases where data are scarce or has long gaps, it is reasonable to apply open datasets and use machine learning techniques like RF and SVM to construct a new data or to fill the rainfall data gaps.

Acknowledgments The authors are thankful to the National Hydrology Project (NIH SP-45) for funding the study and to support this research work. The authors would like to thank the National Institute of Hydrology India for providing facilities to carry out this research work. The authors are also thankful to the Indian Meteorological Department Pune for providing the gridded precipitation dataset. The authors are thankful to WRD Mizoram for providing the observed gauge datasets. The authors are obliged to CHIRPS, TRMM, APHRODITE, and PRINCETON data generation teams/organizations for providing the rainfall datasets at free of cost. The authors are thankful to Python developers project team who made the software/scripts/libraries available free of cost.

References

Chen, C., Zhang, Q., Kashani, M. H., Jun, C., Bateni, S. M., Band, S. S., Dash, S. S., & Chau, K. W. (2022). Forecast of rainfall distribution based on fixed sliding window long short-term memory. *Engineering Applications of Computational Fluid Mechanics, 16*(1), 248–261.

Dwivedi, D. K., Kelaiya, J. H., & Sharma, G. R. (2019). Forecasting monthly rainfall using autoregressive integrated moving average model (ARIMA) and artificial neural network (ANN) model: A case study of Junagadh, Gujarat, India. *Journal of Applied and Natural Science, 11*(1), 35–41.

Funk, C., Peterson, P., Landsfeld, M., Pedreros, D., Verdin, J., Shukla, S., Husak, G., Rowland, J., Harrison, L., Hoell, A., & Michaelsen, J. (2015). The climate hazards infrared precipitation with stations – A new environmental record for monitoring extremes. *Scientific Data, 2*, 150066.

Grömping, U. (2009). Variable importance assessment in regression: Linear regression versus random forest. *American Statistician, 63*(4), 308–319.

He, R., Zhang, L., & Chew, A. W. Z. (2022). Modeling and predicting rainfall time series using seasonal-trend decomposition and machine learning. *Knowledge-Based Systems, 251*, 109125.

Mariette, A., & Rahul, K. (2015). Support vector regression. In *Efficient learning machines* (pp. 67–81).

Misra, R., Kumar Panda, P., and Behera, D., 2020. Rainfall prediction using machine learning approach: A case study for the state of Odisha production of gas hydrates view project DST health care view project.

Simpson, J., Kummerow, C., Tao, W.-K., & Adler, R. F. (1996). *On the tropical rainfall measuring Mission (TRMM)* (Vol. 60, p. 19). Springer.

Singh, V., Jain, S. K., & Singh, P. K. (2019). Inter-comparisons and applicability of CMIP5 GCMs, RCMs and statistically downscaled NEX-GDDP based precipitation in India. *Science of the Total Environment, 697*, 134163.

Svetnik, V., Liaw, A., Tong, C., Christopher Culberson, J., Sheridan, R. P., & Feuston, B. P. (2003). Random Forest: A classification and regression tool for compound classification and QSAR modeling. *Journal of Chemical Information and Computer Sciences, 43*(6), 1947–1958.

Tharun, V. P., Prakash, R., & Devi S. R. (2018). *Proceedings of the International Conference on Inventive Communication and Computational Technologies : ICICCT 2018 : 20–21, April 2018.*

Vapnik, V. (1999). *Statistics for engineering and information*. Springer.

Wagner, P. D., Fiener, P., Wilken, F., Kumar, S., & Schneider, K. (2012). Comparison and evaluation of spatial interpolation schemes for daily rainfall in data scarce regions. *Journal of Hydrology, 464–465*, 388–400.

Wei, W., Yan, Z., & Jones, P. D. (2020). A decision-tree approach to seasonal prediction of extreme precipitation in eastern China. *International Journal of Climatology, 40*(1), 255–272.

Yatagai, A., Kamiguchi, K., Arakawa, O., Hamada, A., Yasutomi, N., & Kitoh, A. (2012). Aphrodite constructing a long-term daily gridded precipitation dataset for Asia based on a dense network of rain gauges. *Bulletin of the American Meteorological Society, 93*(9), 1401–1415.

Zainudin, S., Jasim, D. S., & Bakar, A. A. (2016). Comparative analysis of data mining techniques for malaysian rainfall prediction. *International Journal on Advanced Science, Engineering and Information Technology, 6*(6), 1148–1153.

MIX
Papier aus verantwortungsvollen Quellen
Paper from responsible sources
FSC® C105338
FSC
www.fsc.org

If you have any concerns about our products,
you can contact us on
ProductSafety@springernature.com

In case Publisher is established outside the EU,
the EU authorized representative is:
Springer Nature Customer Service Center GmbH
Europaplatz 3, 69115 Heidelberg, Germany

Printed by Libri Plureos GmbH
in Hamburg, Germany